T0140773

Bibliografische Information der Deutschen Nationalbibliothek

Die Deutsche Nationalbibliothek verzeichnet diese Publikation in der Deutschen Nationalbibliografie; detaillierte bibliografische Daten sind im Internet über http://dnb.d-nb.de abrufbar.

ISBN 978-3-8325-3748-7

Logos Verlag Berlin GmbH
Comeniushof, Gubener Str. 47,
10243 Berlin
Tel.: +49 (0)30 42 85 10 90
Fax: +49 (0)30 42 85 10 92
INTERNET: http://www.logos-verlag.de

Parallel Monte-Carlo Tree Search for HPC Systems and its Application to Computer Go

Dissertation

A thesis submitted to the
Faculty of Electrical Engineering, Computer Science and Mathematics
of the
University of Paderborn
in partial fulfillment of the requirements for the
degree of *Dr. rer. nat.*

by

Lars Schäfers

Paderborn, Germany
March 2014

Supervisors:
Prof. Dr. Marco Platzner
PD Dr. Ulf Lorenz

Reviewers:
Prof. Dr. Marco Platzner
Prof. Dr. Ingo Althöfer
PD Dr. Ulf Lorenz

Additional members of the oral examination committee:
Prof. Dr. Friedhelm Meyer auf der Heide
Dr. Matthias Fischer

Date of submission:
27.03.2014

Date of public examination:
27.05.2014

Acknowledgments

First of all I would like to thank my advisors Prof. Dr. Marco Platzner and PD. Dr. Ulf Lorenz for supporting my research. They created an ideal environment, starting with their efforts in organizing a PhD Scholarship with Microsoft Research and convincing me of doing this PhD research in the area of Computer Go. I have benefited greatly from their experience and advices in the entire course of my PhD studies.

Furthermore I would like to thank:

- Prof. Dr. Ingo Althöfer (Jena), for an hours lasting and highly motivating discussion at the European Go Congress in Bonn Bad Godesberg in 2012. Ingo inspired me to carry out research on the analysis of score histograms that finally resulted in a Best Paper Award and Chapter 6 of this thesis.
- Prof. Dr. Friedhelm Meyer auf der Heide and Dr. Matthias Fischer, for serving on my oral examination committee.
- Microsoft Research Cambridge for granting me a Phd Scholarship as a funding for my first three years of PhD research. Special thanks to Thore Graepel, David Stern and Satnam Singh, all researchers at Microsoft Research Cambridge, for their support and valuable discussions.
- Tobias Kenter, my longtime office colleague, for extensive discussions and valuable support in various respects.
- Axel Keller, for his unbeatable technical and interpersonal support.
- My former student Tobias Graf, for longtime support of my research by writing his Bachelor and Master thesis with me and for having worked as a student assistant on further topics related to MCTS.
- My colleagues Andreas Agne, Jahanzeb Anwer, Tobias Beisel, Alexander Boschmann, Stephanie Drzevitzky, Heiner Giefers, Mariusz Grad, Markus Happe, Tobias Graf, Nam Ho, Server Kasap, Paul Kaufmann, Achim Lösch,

Enno Lübbers, Sebastian Meisner, Björn Meyer, Christian Plessl, Heinrich Riebler, Tobias Schumacher, Gavin Vaz and Tobias Wiersema for valuable discussions.

- Bernard Bauer for his kind organizational support from the first day on.
- Jens Simon and all the members of the Paderborn Center of Parallel Computing for numerous valuable discussions.
- Klaus Petri, one of Germany's strongest Go players, for extensive testing of our Go engine Gomorra and a huge number of suggestions and valuable hints for further improving Gomorra's playing strength.
- Shih-Chieh (Aja) Huang, author of the Go playing program Erica and winner of a gold medal at the prestigious ICGA Computer Olympiad in Kanazawa (Japan) in 2010, for many inspiring discussions and helpful hints on becoming stronger on the 19×19 board.
- Andreas Fecke, Martin Hershoff and the entire Paderborn Go Club, for teaching me the game of Go. I know, I was a rather weak student but I'm proud to see Gomorra becoming one of our strongest club members.
- The Bachelor and Master students I have supervised for helping me implement and supporting my research.
- The staff of the Cologne compute center Regionales Rechenzentrum der Universität Köln (RRZK) for granting me access to the CHEOPS cluster. A huge number of early experiments with our MCTS parallelization were performed on the CHEOPS HPC cluster. I especially thank Stefan Borowski and Viktor Achter for their kind support and numerous helpful comments on my work.
- The members of the Computer-Go mailing list and the entire Computer Go community for valuable discussions and the motivating and kind atmosphere in our regular Go tournaments. Many thanks to Nick Wedd for kindly organizing such tournaments since May 2005.

Finally, I would like to thank my parents and my brother and sister for their support. I thank my wife Anne, sweet Greta and little Matheo. Seeing you smile when coming home is the greatest thing in life.

Abstract

Monte-Carlo Tree Search (MCTS) is a class of simulation-based search algorithms. It brought about great success in the past few years regarding the evaluation of deterministic two-player games such as the Asian board game Go. A breakthrough was achieved in 2006, when Rémi Coulom placed 1st at the ICGA Computer Go Olympiad in Turin with his MCTS based Go programm CrazyStone in the 9×9 devision. Until today, MCTS highly dominates over traditional methods such as $\alpha\beta$ search in the field of Computer Go.

Over the years, MCTS found applications in several search domains. A recent survey of MCTS methods lists almost 250 MCTS related publications originating only from the last seven years, which demonstrates the popularity and importance of MCTS. It is currently emerging as a powerful tree search algorithm yielding promising results in many search domains such as connection games Hex and Havannah, combinatorial games Breakthrough and Amazons as well as General Game Playing and real-time games. Apart from games, MCTS finds applications in combinatorial optimization, constraint satisfaction, scheduling problems, sample-based planning and procedural content generation.

In this thesis, we present a parallelization of the most popular MCTS variant for large HPC compute clusters that efficiently shares a single game tree representation in a distributed memory environment and scales up to 128 compute nodes and 2048 cores. It is hereby one of the most powerful MCTS parallelizations to date. We empirically confirmed its performance with extensive experiments and showed our parallelization's power in numerous competitions with solutions of other research teams around the world.

In order to measure the impact of our parallelization on the search quality and remain comparable to the most advanced MCTS implementations to date, we implemented it in a state-of-the-art Go engine *Gomorra*, making it competitive with the strongest Go programs in the world.

Apart from the parallelization, we present an empirical comparison of different Bayesian ranking systems when being used for predicting expert moves for the game of Go. The ranking systems are based on the training of parametrized probability models. Those models make use of shape patterns and tactical feature patterns to abstract pairs of a position and a move that can be played in that position. Adapting the parameters of such prediction systems during an ongoing MCTS search is considered a promising direction for further search quality improvement.

We investigated the automated detection and analysis of evaluation uncertainties that show up during MCTS searches. This was done with the objective to develop triggering mechanisms for guiding the aforementioned adaptation of move prediction systems. As the result, we obtained a promising system that is capable of detecting ongoing local fights in Go positions in the course of an MCTS search that even allows for approximately locating involved regions on the Go board.

Zusammenfassung

Monte-Carlo Tree Search (MCTS) beschreibt eine Klasse von simulationsbasierten Baumsuchalgorithmen. In den vergangenen Jahren wurden mit ihr enorme Fortschritte in der Bewertung von Zweipersonenspielen wie dem asiatischen Brettspiel Go erzielt. Ein Durchbruch wurde im Jahr 2006 erzielt, in dem Rémi Coulom mit seinem MCTS basierten Go Programm CrazyStone im 9×9 Go den Sieg bei der ICGA Computer Go Olympiade in Turin erringen konnte. Bis heute dominieren MCTS Algorithmen die Entwicklung im Computer Go deutlich vor traditionellen Methoden wie der $\alpha\beta$ Suche.

Über die Jahre haben MCTS Algorithmen Anwendung in diversen Suchdomänen gefunden. Eine kürzlich veröffentlichte Studie führt über 250 Publikationen auf, die in den vergangenen sieben Jahren im Bereich MCTS veröffentlicht wurden und demonstriert damit die aktuelle Popularität und Relevanz des Verfahrens. MCTS präsentiert sich zunehmend als Klasse leistungsstarker Baumsuchalgorithmen die zu beachtlichen Ergebnissen in vielen Suchdomänen führen. Zu diesen Domänen zählen Spiele wie Hex und Havannah, die kombinatorischen Spiele Breakthrough und Amazons wie auch das General Game Playing und Echtzeit Spiele. Neben Spielen findet MCTS Anwendung in kombinatorischer Optimierung, Erfüllbarkeitsproblemen, Scheduling Problemen, samplebasierter Planung und prozeduraler Generierung.

In dieser Arbeit stellen wir für die derzeit populärste MCTS Variante eine Parallelisierung für große HPC Cluster vor. Unsere Parallelisierung hält eine einzelne Suchbaumrepräsentation in verteiltem Speicher vor und skaliert für bis zu 128 Rechenknoten und 2048 Cores. Heute ist sie damit eine der leistungsstärksten MCTS Parallelisierungen. Wir haben die Leistungsfähigkeit unserer Parallelisierung mit umfangreichen Experimenten empirisch belegt. Zusätzliche empirische Vergleiche mit Lösungen anderer Forschergruppen bestätigen die Qualität unseres Verfahrens.

Um die Auswirkungen unserer Parallelisierung auf die Qualität der Suche bestimmen zu können, haben wir diese für unser state-of-the-art Go Programm *Gomorra* implementiert. Dies erlaubt einen Vergleich mit den aktuell fortgeschrittensten MCTS Implementierungen anderer Forschergruppen. So konnten wir in internationalen Wettbewerben zeigen, dass sich unser paralleles Go Programm auf Augenhöhe mit den weltweit stärksten Programmen Anderer befindet.

Neben unserer Parallelisierung präsentieren wir einen empirischen Vergleich verschiedener Bayesianischer Zugvorhersagesysteme bei ihrer Anwendung im Computer Go. Diese Systeme basieren auf dem automatisierten Training von parametrisierten Wahrscheinlichkeitsmodellen. Die Modelle nutzen Muster aus Steinformationen und taktischen Eigenschaften um Paare von Spielpositionen und verfügbaren Spielzügen zu abstrahieren. Die Anpassung der Parameter solcher Vorhersagesysteme während einer MCTS Suche wird gemeinhin als vielversprechende Forschungsrichtung angesehen.

Wir haben zudem die automatisierte Erkennung und Analyse von Bewertungsunsicherheiten, die während MCTS Suchen auftreten untersucht. Die Zielsetzung war die Entwicklung von Mechanismen zur gezielten Steuerung der vorgenannten Parameteranpassung von Vorhersagesystemen zur Laufzeit. Das Ergebnis ist ein vielversprechendes Verfahren, das erlaubt, andauernde lokale Kämpfe in Go Positionen zu erkennen und auf dem Spielbrett zu lokalisieren.

Contents

List of Figures

CHAPTER 1

Introduction

1.1 Motivation

The game of Go is said to fascinate people for more than 4000 years. The oldest known written recording in history dates back to the year 625 B.C. [10] and today, the total number of active Go players is estimated to be between 25 and 50 million worldwide. In 1710, Gottfried Wilhelm Leibniz mentioned the game with the following words: "The multitude of pebbles & size of the gameboard makes, as I should believe easily, this game to have greatest cleverness and difficulty[...]" [63] (original source in latin). Indeed, only recently (in relation to the before mentioned dates), Lichtenstein and Sipser showed the PSPACE hardness of the generalized game of Go for arbitrary board sizes from todays computational comlexity theory's perspective [65]. It is hereby not surprising, that until today, no computer program exists that is able to compete with the current strongest human Go players. And this holds true, although research in Computer Go is carried out for decades already. At the latest in 1985, with the proclamation of the Ing-Prize, a 40.000.000 NT dollar[1] prize sponsored by Ing Chang-ki for the first program to beat a professional Go player in four out of seven even games, Computer Go became a popular subject for research. It was often cited as a grand challenge for artificial intelligence.

In 2006, the rise of Monte Carlo Tree Search (MCTS), a simulation based search method, revolutionized the development of Computer Go programs. Within only seven years, Go programs got to the strength of strong amateur players, reducing the gap to the strongest human professional Go players significantly. Besides

[1] 40.000.000 New Taiwan dollar had a worth of about 1.000.000 USD at that time

constantly improving techniques for learning and predicting expert play and the continuous development of MCTS enhancements, an almost safe source for further strength improvement is the sole increment of the number of simulations that are computed per move decision. But MCTS also requires us to keep a search tree representation in memory. Hence, an increased number of simulations also demands for more memory to store an ever growing search tree representation. This motivates the use of large compute clusters, that not only provide large amounts of computation cores that are needed to increase the number of simulations computable per time unit, but at the same time, provide a huge amount of memory capable of storing large search tree representations. Consequently, a central part of this thesis copes with the parallelization of MCTS for large compute clusters.

Although a lot of strength improvement comes from the mere increment of the number of simulations computable per time unit, there remain situations where MCTS in its most basic form appears inappropriate. For example so called capturing races, that require deep lines of correct play. While a basic MCTS searcher needs to rediscover those lines in different subtrees of the overall search tree, humans are more capable to exploit the locality of such problems and succeed to break down whole positions into smaller subgames where appropriate. Humans thereby manage to share information between different subtrees of the overall search tree. We expect an enormous improvement of search quality by a successful extension of MCTS with comparable capabilities.

1.2 Contributions of This Thesis

The main contribution of this thesis is the development of a new technique for scalable parallel Monte-Carlo Tree Search on high performance computing systems (HPC). Our parallelization's unique feature is a single large game tree representation, distributed among the local memories of distinct compute cluster nodes. We furthermore contribute to the long term objective of dynamically adapting simulation policies based on generalizing and sharing obtained information between different subtrees of a single large game tree. More detailed, we made the following contributions to the research field of Monte-Carlo Tree Search:

- We developed a novel technique for distributed Monte Carlo Tree Search based on a data driven approach, that we called *Distributed-Tree-Parallelization*. For this purpose, we implemented a highly efficient distributed hash table, that is able to store a large game tree, distributed on the memories of distinct compute nodes. Our parallelization hereby excels in regard of the realizable size of the search tree representation in memory and consequently allows for better exploitation of obtained simulation results, directly leading to a more informed search. A *data driven approach* here denotes that the distribution of computational tasks is based on the location of the data the computations depend on. Hence, by prescribing the distribution of the game tree data structure among the single compute nodes' local memories, we implicitly determine a policy for assigning computational tasks, that depend on the game tree data, to particular compute nodes of a cluster. We concentrate on compute clusters that are made of homogeneous compute nodes, each being equipped with potentially more than one many-core CPU. We assume a modern low latency interconnect of the compute nodes, e.g., an Infiniband network. With an implementation of Distributed-Tree-Parallelization in our Go playing program Gomorra, we empirically showed its scaling to more than 2000 compute cores in the best setting.

- We developed the before mentioned MCTS based Go program Gomorra and parallelized it for shared memory machines, as well as for large HPC compute cluster systems. By integrating a large number of modern heuristics and techniques, we made it comparable to the current strongest Go programs in the world. This strengthens our firm conviction that our findings and improvements are relevant and for most parts also applicable to other state-of-the-art MCTS searchers. We proved the strength of Gomorra in several international computer Go tournaments, most recently in 2013 by winning a silver medal at the 13th Computer Olympiad in Yokohama, Japan.
- One of the central sources for search quality improvement, not only for MCTS, is the development of high quality move prediction systems. Due to their relevance, a number of move prediction systems were developed and investigated for computer Go. For the first time, we compared several bayesian move prediction systems under fair and equal conditions to reveal their efficiency when being faced with identical data records. We showed that, given equal environments, most of the prediction systems under review perform almost identical. Some of those systems however, are filtering algorithms, i.e., they can learn from data streams. Thereby, they appear ideal for being used to train or adapt prediction models during actual MCTS search runs, by processing simulation results as they occur.
- We believe that, one of the most promising future directions for further search quality improvement in MCTS, will be the dynamical adaptation of playout policies based on the outcome of former simulations. While MCTS shows great performance in a wide range of Go positions, there remain certain situations where MCTS fails. For example lengthy capturing races, especially when they show up simultaneously in a multitude. We developed a method that continuously analyses histograms of simulation outcomes to discover capturing races and further difficult situations that require special attention during MCTS search runs. Furthermore, once the histogram analysis signals

a difficult to handle situation, we provide a method to heuristically localize regions on the Go board that are responsible for the observed singularities. For the future, this might allow a special treatment of those regions with the aim to more correctly estimate the game positions' values and thereby might lead to substantial search quality improvements.

1.3 Thesis Structure

The remainder of this thesis is structured as follows:

Chapter 2 gives a brief introduction to the game of Go, game tree search and the history of Computer Go.

Chapter 3 gives a detailed explanation of the general Monte Carlo Tree Search framework and presents a number of extensions and variants developed throughout the past seven years.

Chapter 4 is devoted to the central contribution of this thesis and presents our proposal for parallelizing MCTS on hybrid shared and distributed memory systems. It furthermore summarizes related work on MCTS and attempts on its parallelization.

Chapter 5 presents an empiric comparison of several Bayesian prediction systems when being used as move predictors for the game of Go.

Chapter 6 is dedicated to the presentation of a method for automated analysis of score histograms obtained from MCTS simulations. We present a method for heuristically detecting and localizing evaluation uncertainties during MCTS searches.

Chapter 7 summarizes and concludes the work presented in this thesis. It furthermore lists a number of promising future directions.

CHAPTER 2

Background and Related Work

2.1 The Game of Go

One of the interesting properties of the game of Go, despite its age and history, is the small set of rules and the rules' simplicity. Nevertheless, deciding for the best move in an arbitrary Go position turns out to be a highly complex task. These properties made Go become a popular subject to a number of mathematical investigations, some of them dating back to the 11th century. This section introduces the basic rules of the game and briefly summarizes investigations on the complexity of the game. Due to the long history of the game of Go, a number of slightly different rule sets evolved. Two of the most popular are known as the Japanese and Chinese rules. Throughout this thesis, we use solely a variant of the Chinese rules and therefore we stick to a description of the basic Chinese rules only. For a more detailed discussion on the various Go rule sets and their differences, we refer the interested reader to the British Go Association's website: `http://www.britgo.org/rules/compare`.

2.1.1 Rules and Terms

The game of Go is a two-player, zero-sum game. The zero-sum property here says, that an advantage for one player is to the exactly same amount a disadvantage for the other player. Go is played on a game board called *Goban*. It consists of 19 horizontal and 19 vertical lines, that intersect in 361 points. Figure 2.1 shows an empty Goban, that constitutes also the starting position of each game. Nine intersections are highlighted by little dots that are called *hoshi*. They are mainly intended to assist humans' orientation. Both players are given a number of stones. One player has solely black stones, while all stones of the other player are white. Accordingly, we distinguish between the black player and the white player. We simply call them Black and White in the remainder. Both players alternate in placing a single stone of their color on one of the not yet occupied 361 intersections. Black always plays the first stone in a game. We call horizontally and vertically neighboring stones of equal color *groups*. A group might also consist of a single stone. Figure 2.2a shows an example position containing eight groups of stones that are accordingly numbered. Stones with equal numbers belong to the same group.

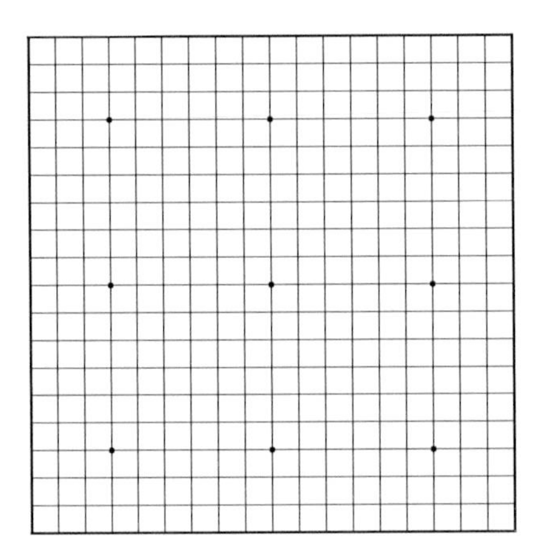

Figure 2.1: The empty Goban

Liberties and Capturing

Once a stone is placed on the board, it is never moved again. The only exception to this rule is for stones that get *captured*. We call horizontally or vertically adjacent empty crossings of groups *liberties*. In case the number of liberties of a group gets reduced to zero, the group is said to be captured and all stones belonging to it, are removed from the board. In the position depicted in Figure 2.2b, three intersections are labeled with letters. They mark last liberties of groups. Note that a is the last liberty of two different groups, i.e., group 1 and 3, at the same time. Hence, placing a white stone at a, as shown in Figure 2.2c, reduces the number of liberties of groups 1 and 3 to zero, leading to a capture of both groups. All stones of the captured groups are then removed from the board and become prisoners of the white player. Figure 2.2d shows the resulting position.

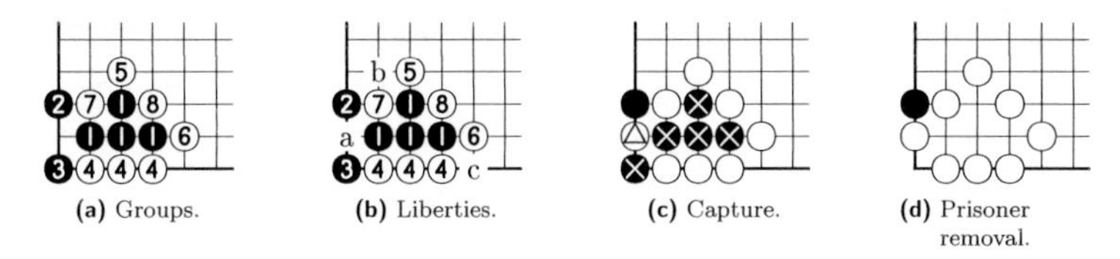

(a) Groups. (b) Liberties. (c) Capture. (d) Prisoner removal.

Figure 2.2: Terms of the game of Go (part 1)

Suicide, Eyes and Unconditional Life

Following the rules described so far, it appears possible to place a stone on the board that immediately stays without liberties, and hence would become a prisoner of the opponent. An example is shown in Figure 2.3a, where the white stone marked with a triangle was played last. Such moves are called *suicide* moves and are generally forbidden[1]. As can be seen in Figure 2.3b, the move discussed before is valid, if Black plays in its other liberty first. In this case, again, the white stone seems to stay without liberties when being placed on the board, but this is also true for the adjacent black group. The captured stones of the opponent, in this case the black group, are removed first. Hence, a single white stone with three liberties will validly remain on the board.

With the observations made before, one can infer another important property of the black groups in the position depicted in Figure 2.3c. There is no way for White to play on any of Black's liberties labeled with the letter *a*, if Black will not do so before. Hence, Black's group can not be captured by White. We call Black's group *unconditionally alive* and a liberty like *a*, that is completely surrounded by stones of the same color, *eye*. An eye does not need to be a single crossing, but might be a number of adjacent empty crossings. One can show that the essential requirement for a group of stones to be unconditionally alive is its containment of at least two eyes.

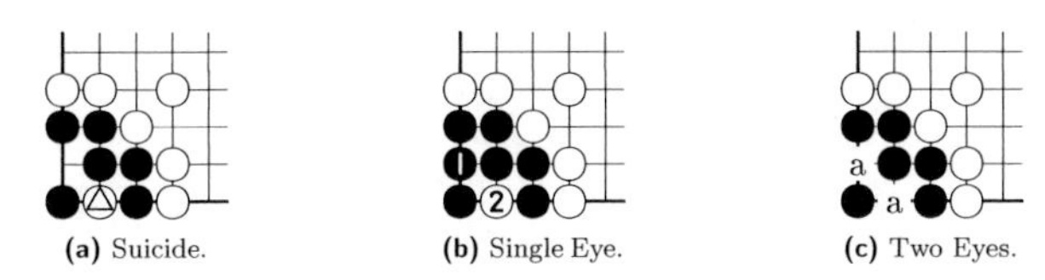

(a) Suicide. (b) Single Eye. (c) Two Eyes.

Figure 2.3: Terms of the game of Go (part 2)

[1]Actually some rule sets, like the New Zealand rules, allow suicide moves.

Game Ending, Scoring and Seki

Besides placing a stone on the board, a player always has the option to pass, i.e., to hand over the right of placing a stone to the opponent without placing a stone himself. The game ends when both players pass in a row[2]. Thereafter, *dead* stones are removed from the board. Stones are considered dead when they are not able to form living groups or connect to such. If players do not agree on the life-dead status of groups, they might resume play to clarify the situation. Figure 2.4a shows an example 7×7 Go position with a single dead black stone, marked with a square. Hence, after removal of all dead stones, the position in Figure 2.4b remains for scoring.

The board can now be divided into black and white *areas*. A field is considered black area if it is occupied by a black stone, or if it is part of a group of empty fields that is adjacent solely to black stones or an edge of the board. The white area is determined accordingly. The set of intersections that are part of a players area but not occupied by a players stone (i.e., they are empty) is also called the player's territory. Figure 2.4c shows an example of a terminal game position. Fields that correspond to Black's area are labeled with the letter b, White's area is marked with w. Simple colorwise counting of the number of stones and empty area intersections yield the scores of each player. In the example position, we have $20 + 3 = 23$ points for Black and $20 + 4 = 24$ points for White. Hence, from the sole interpretation of the board configuration, White wins by one point.

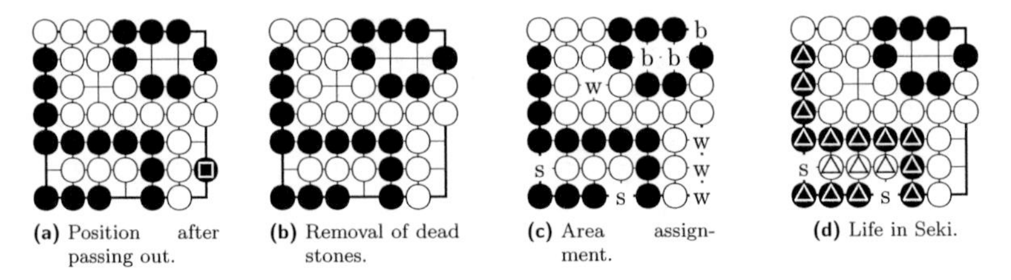

(a) Position after passing out. (b) Removal of dead stones. (c) Area assignment. (d) Life in Seki.

Figure 2.4: Scoring example on a 7×7 Go position

But Figure 2.4c also shows two points that are marked with an s, indicating that those points are neither white nor black area. Looking at the adjacent groups marked with a triangle in Figure 2.4d, one might argue that at least some of those stones should be considered as dead, as definitely, not all of them are able to form two eyes or connect to other living groups. Whichever player first places a stone on one of the intersections labeled with the letter s thereby reduces the number

[2]It is also possible that one of the players resigns, accepting the other player as the winner. This is also a common way of finishing a game, especially between strong players that are able to predict a game's outcome at an early stage. Resigning early in a clearly decided game is also considered as a matter of politeness.

of liberties of its own groups to one and will be captured by the opponent during the next turn. Accordingly, none of the player wants to play on s. The involved groups are denoted as *alive in seki* and remain on the board.

Ko rules

Figure 2.5 shows a simple, two moves lasting sequence that leads to a position repetition. By placing a stone on the intersection labeled with the letter a in Figure 2.5a, White can capture a single black stone, leading to the position depicted in Figure 2.5b. Now Black can capture the newly placed white stone, creating the position shown in Figure 2.5c that equals the initial position.

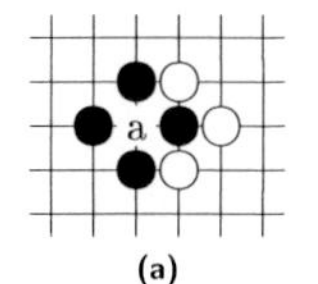

(a)

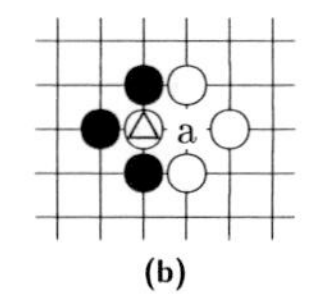

(b)

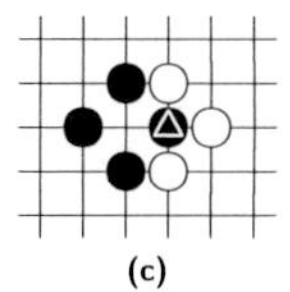

(c)

Figure 2.5: Basic Ko.

Hence, given the rules as explained so far, repetitions of positions, and thereby never ending games are possible. The so-called *ko* rules target at prohibiting those position repetitions. While a number of ko rules do exist and are regularly subject of lengthy discussions, we will present and discuss only the so-called *positional super ko* (PSK) rule, that is the only ko rule used throughout this thesis. PSK simply forbids the repetition of complete board configurations, regardless of which player is to move[3]. The pass rule obviously takes a special role, otherwise passing would not be allowed, because the board configuration remains the same. Hence, what was said before remains true: for either player, passing is always a valid option.

Semeai

In many respects, Go is a game about counting. This is especially true in the manner of counting liberties, when two or more adjacent groups take part in a so-called *capturing race*. Two groups take part in a capturing race if they both need to capture the respective opponent group in order to create two eyes. The difference in the number of liberties is often critical for the outcome of such races. The Japanese term for capturing race is *semeai*. Semeai play an important role in Computer Go and are discussed in more detail in Chapter 6.

[3] A corresponding, player sensitive super ko rule is called *situational super ko*.

Player Ranking, Handicapping and Komi

Go players are traditionally ranked with a system of *kyu* and *dan* ranks. Hereby, the kyu ranks are considered as *student* ranks that are counted downwards with increasing strength, ranging from 30 kyu for the very beginner to 1 kyu for the strongest students. The dan ranks are devoted to *masters* and are counted upwards, starting with the 1st and ending with the 7th dan. Additional ranks, typically called *professional dan* are reserved for professional Go players. They are awarded by the professional Go associations of some Asian countries, currently Japan, China, Korea and Taiwan. The ranks range from 1 to 9 professional dan, where players holding a professional dan are generally expected to be stronger than amateur players. Hence, a holder of the 1st professional dan might be stronger than the holder of a 7th amateur dan. A difference of one amateur rank, either kyu or dan, is expected to be of the value of one move. That is, a game between a 6th and 3rd kyu can be expected to be even, when the 6 kyu player is allowed to make the first three moves in the game before the stronger 3 kyu player joins the game. The practice of giving the weaker player the right of placing a number of stones according to the ranking difference to his opponent before the actual game starts is called *handicapping*. Another way of even out unequal games is the concept of *komi*. Komi is a certain number of points awarded to the white player even before the game starts, to compensate for the advantage of the black player for moving first. An adequate komi depends on the rank difference of the players and the size of the Go board. Typical komi values for the 19×19 board and equally ranked players are 6.5 and 7.5. Choosing the komi non-integral prevents the occurrence of draws. The correct komi[4] for 9×9 Go is currently assumed to be 7.0.

Relative Elo Rating

Another popular system for rating players is the Elo system. It is named after its inventor Arpad Elo [34] and was originally developed for the rating of chess players. A player's rating is given in form of a single rating number representing the player's skill. The Elo system is based on the simplifying assumption that each player's performance can be modeled with a normally distributed random variable of equal standard deviation. The mean of this distribution is taken as the player's skill, or Elo rating. In the remainder of the thesis, we use the *relative Elo* measure, when presenting results of empirical experiments with different Go engines. Given two players, or Go playing programs, A and B, with corresponding absolute Elo ratings R_A and R_B, respectively, we obtain the relative Elo of A over B by $\text{Elo}_{\text{Rel}}(A, B) := R_A - R_B$. Given $\text{Elo}_{\text{Rel}}(A, B)$ we can estimate the probability

[4]The *correct* komi targets on exact compensation of Black's advantage of playing first assuming optimal play by both players.

$P_{A,B}$ of A to win a game when competing with B, based on the definition of the Elo rating system, by

$$P_{A,B} := \frac{1}{1 + 10^{-\mathrm{Elo}_{\mathrm{Rel}}(A,B)}/400} \ .$$

Accordingly, given the probability of $P_{A,B}$ of A to win a game, when competing with B as an observation of a number of games between A and B, we can infer the corresponding relative Elo rating by

$$\mathrm{Elo}_{\mathrm{Rel}}(A,B) := 400 \log_{10}\left(\frac{1}{(1 - P_{A,B})} - 1\right) \ .$$

Table 2.1 shows some example mappings of relative Elo points to the corresponding probabilities of winning future games.

Table 2.1: Mapping of relative Elo points to probabilities of winning.

$\mathrm{Elo}_{\mathrm{Rel}}(A,B)$	$P_{A,B}$
0	0.5
10	0.5144
50	0.5715
100	0.6401
200	0.7597
300	0.8490
400	0.9091
500	0.9468
600	0.9693

2.1.2 Complexity

Given an arbitrary $n \times n$ Go position, the computational complexity of determining the winner assuming optimal play, is *polynomial space hard* (PSPACE-hard). This was shown in 1980 by Lichtenstein and Sipser [65], by reducing the canonical PSPACE-complete problem TQBF to the game generalized geography, then to a planar version of generalized geography and finally to the game of Go. It is thereby shown that the game of Go, generalized in respect to the board size, is at least as difficult as the most difficult problems in the complexity class PSPACE. It would be considered as PSPACE complete, if the length of the game could be bounded polynomially in the number of intersections of the board. In 1983, Robson [82] showed that Go is EXPTIME-complete for certain ko rules.

Considering computational complexity only, due to its constant problem size, 19×19 Go is decidable in polynomial time. Using a large lookup table that contains the optimal move to each position would bound the computation time to $O(1)$. Hence, for discussing the complexity of a practical board size of 19×19, we are also interested in the value of the constants that describe the problem, like the number of legal positions the game might enter, i.e., the state-space complexity, the number of possible games and further parameters of the game tree like its average branching factor. In 2007, Tromp and Farnebäck [100] published their results about investigations on the numbers of legal positions for varying board sizes. They computed the exact numbers for board sizes of up to 17x17 and approximated the number of legal 19×19 Go positions to be pretty exactly $2.08168199382 \cdot 10^{170}$. Here, *legal* does not necessarily mean that all positions can occur during a valid game, but only that all groups of stones on the board have at least one liberty. Tromp and Farnebäck also state, a lower and upper bound on the number of possible 19×19 Go games to be $10^{10^{48}}$ and $10^{10^{171}}$ respectively.

The typical length of a 19×19 Go game between strong players is between 150 and 250 moves with an average number of choices of about 200 to 250. But the vast amount of legal positions and games and the high number of choices a player has for each move alone doesn't necessarily make the game particularly complex. It might still be the case that a relatively easy policy exists that leads to optimal play. However, no such policy is known and after decades of research that was carried out on computer Go, no solution was found to estimate the value of arbitrary Go positions good enough to safely reduce the search space. This additional fact makes the search for the best move to arbitrary Go positions an highly challenging task. The proven PSPACE-hardness of generalized Go makes it unlikely that a strong and computationally efficient static evaluation function for arbitrary Go positions will be discovered in the future.

2.2 Game Tree Search

The objective of our research on a Go playing program is the computation of the best possible move to any arbitrary Go position. Given this objective, we have to define how the *best* move is characterized. Thereafter, in this section, we will discuss how its computation can be realized.

Given an arbitrary Go position and a player to move, there typically exist a number of possible choices for the next move. Whichever move the player chooses, there will be a resulting new position. Writing down both these positions, connecting them with an arrow representing the move and doing so recursively for all moves and resulting positions, we end up with a drawing like the one shown in Figure 2.6. In the Figure, a little dot on the stones indicates the stone that was placed last. This procedure of writing down all possible continuations of the game, starting with the position on the top, leads us to the idea of the so-called *game tree*.

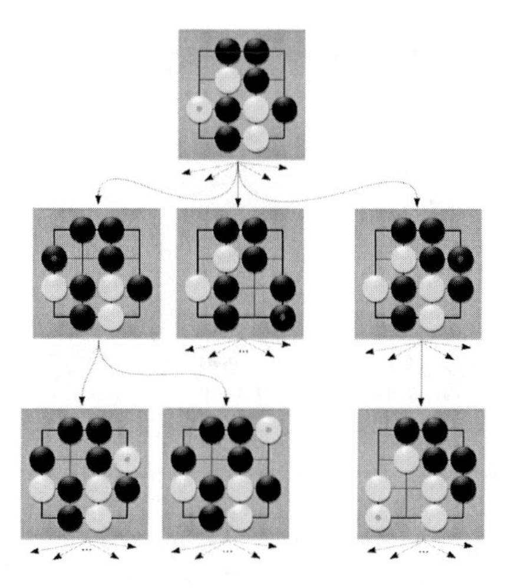

Figure 2.6: Example game tree for 4×4 Go.

2.2.1 Terminology

We will now introduce a number of formalisms that lead us to a formal description of a game tree. This gives us the background to define the *best* move and allows for discussing its computation.

Definition 1 (Directed Acyclic Graph (DAG)) *A directed graph $G := (V, E)$ is a 2-tuple of a set of nodes V and a set of directed edges $E \subseteq (V \times V)$, each connecting two nodes. We call a sequence of nodes $P := (v_1, v_2, \ldots, v_n)$ with $n > 1$ a path in G, iff $(v_i, v_{i+1}) \in E$ for all $1 \leq i < n$. P is called a cycle, iff $v_1 = v_n$. We call a graph a directed acyclic graph (DAG) iff it contains no cycles.*

Definition 2 (Tree) *We call a graph $G := (V, E)$ tree, if it is a DAG and if there is exactly one node v_0 with no incoming edge, i.e., $(v_j, v_0) \notin E$ for all j. All other nodes v_i have a single incoming edge, i.e., there exists exactly one j with $(v_j, v_i) \in E$. We call v_0 the root of the tree. We call the set $L := \{v_l | \forall_j (v_l, v_j) \notin E\}$ of nodes with no outgoing edges the set of leaves of the tree.*

Definition 3 (Game Tree) *We define a game tree $G := (V, E, r)$ as a tree with an additional leaf value function $r : L \to \mathbb{Z}$. Nodes then represent positions of a game and edges the corresponding moves that lead from one position to the next. Leaf nodes represent terminal positions of the game. The leaf value function r assigns some final score to each terminal game position.*

We now have a formal definition of the game tree that can be used to represent all possible continuations of a game. When selecting the initial position of a game as the root, a game tree contains the entire set of positions the game might enter and even more, as some positions might be reached through different move sequences (cf. Section 3.2.3). In Section 2.1.2, we mentioned that the number of positions for 19×19 Go is expected to be a 171 digit decimal.

In two player games like Go, we can group the game tree nodes according to the player that is to move in each position. We call the two resulting sets MIN and MAX and call the according players the MIN player and the MAX player respectively. We define the leaf value function r to assign a negative score to each terminal position that yields a win for the MIN player, and a positive score to winning positions of the MAX player. We can now recursively define the MinMax value function R for each node of a game tree.

Definition 4 (MinMax Value) *Given a game tree $G := (V, E, r)$ of a two-player zero-sum game, with the set of leaf nodes L, the set of MAX nodes and the set of MIN nodes, we define the MinMax value function $R : V \to \mathbb{Z}$ recursively by:*

$$R(v) = \begin{cases} r(v) & \text{if } v \in L \\ \max_{\{u|(v,u) \in E\}} R(u) & \text{if } v \in \text{MAX} \\ \min_{\{u|(v,u) \in E\}} R(u) & \text{if } v \in \text{MIN} \end{cases}$$

The MinMax value at any node is exactly the final score of a game that starts at the nodes' corresponding position, assuming optimal play by both players. This leads us to the definition of a best move for arbitrary positions.

Definition 5 (Best Move) *Given a game tree $G := (V, E, r)$ with root node v_0 and its corresponding MinMax value function R, we define the set of edges that represent the best moves at position v_0 as $B := \{(v_0, v_i) \in E | R(v_0) = R(v_i)\}$.*

As we now have a formal definition of a best move to a given game, we will briefly discuss how the computation of a best move given the complete game tree information is traditionally realized.

2.2.2 Traditional Game Tree Search Algorithms

For some trivial games, it might be the case that a simple and obvious policy exists that leads to optimal play. For games where such policies are less obvious and therefore not known, we can resort to searching the space of possible game continuations to determine the best moves to any position by computing the game tree nodes minmax values. Due to the PSPACE-hardness of generalized $N \times N$ Go, it is pretty certain that no simple and obvious optimal policy exists for Go. The definitions of the MinMax value function and the definition of the best move leads us to a basic game tree search algorithm, called *minmax*.

To compute the set of best moves, given a complete game tree of a two-player zero-sum game, we can compute the minmax value function recursively, starting at the leaf nodes and finishing at the root node. We then only have to compare the direct root successors minmax values with the roots' minmax value to obtain the set of best moves according to Definition 5. The described search algorithm is called *minmax*-Search. We hereby exhaustively search the whole game tree, making the approach a so-called *brute-force* search.

Investigations on the minmax search lead to more efficient variants, the perhaps most notable among them is the $\alpha\beta$ algorithm. The $\alpha\beta$ algorithm reduces the number of nodes of a game tree that need to be searched remarkably. In 1975, Knuth and Moore [59] proofed that, given a game tree with a constant branching factor of b and a depth of t, in the best case, the $\alpha\beta$ algorithm needs to visit as few as $b^{\lfloor t/2 \rfloor} + b^{\lceil t/2 \rceil} - 1$ leaf nodes only. During the search process, at some points, the $\alpha\beta$ algorithm allows to cut off entire subtrees that were proofed to be irrelevant for the minmax value computation due to the insight obtained by the search so far. The order in that game tree nodes are visited by the search, however, plays a key role for the amount of game tree nodes that can be omitted by the algorithm. Being able to safely predict good moves allows for getting near to the before mentioned lower bound of leaf nodes. Hence, the development of move prediction systems becomes important.

There are more reasons to investigate the use of heuristics in game tree search, despite its use for move sorting. Although, with the $\alpha\beta$ algorithm, the search overhead can be reduced significantly, the remaining search space for games like Chess and Go is several orders of magnitude too large to become computable on current and future computing machines[5]. A technique, already mentioned in 1949 by Claude Shannon [90] for the game of Chess, is to artificially limit the search depth and heuristically evaluate the positions that represent the new, artificial leafs of the depth-reduced game tree. The use of limited depth $\alpha\beta$ search with several further enhancements brought enormous success in computer chess and numerous other games. As a general, not surprising rule of thumb, it is the case that deeper searches lead to better estimated minmax values. An apparently rare exception to this rule are so-called pathologies in game trees, where deeper search leads to worse estimates [88][3][71][70][30].

[5]The number of legal 19×19 Go positions is several orders of magnitude higher than the estimated number of atoms in the universe.

The use of depth-limited $\alpha\beta$ search is most efficient if good enough and fast to compute evaluation functions for arbitrary game positions exist. Most of the time, the whole tree search bases on outcomes of heuristic position evaluations instead of exact scores of terminal game positions. Hence, the quality of the evaluation function is of high importance. Any game tree node's minmax value, and accordingly also the decision about the best move, might change, if only a single leaf node's evaluation changes. With the objective to make the search process less prone to evaluation inaccuracies, a search process called *Conspiracy Number Search* was proposed by McAllester [74] in 1988. It focuses on ensuring that the root nodes minmax value can only change if at least c leaf nodes change their evaluation. c is then called *Conspiracy Number*. Further enhancements lead to a more focused search technique called *Controlled Conspiracy Number Search* that was investigated in 2000 by Lorenz [68][69].

2.3 Computer Go

Research on Computer Go has a long tradition, reaching back to the 1960s. In this section, we will give a brief overview of developments in Computer Go throughout the past 50 years and point out the central role of Computer Go research on the recent development of Monte-Carlo Tree Search.

2.3.1 History

According to [15], it seems the first Go program was developed in 1960 by D. Lefkovitz [62] and the first scientific paper was published in 1963 by H. Remus [80]. In 1969, Albert L. Zobrist developed the apparently first Go program that was able to defeat a human player[6] [108][110]. The same Zobrist also published a very popular hashing method for the game of Go that, given the hash value of a Go position, allows for an efficient generation of all hash-values of the corresponding rotated, mirrored and color inverted positions by computations on the hash-value only. The method became very popular and is widely used and known as Zobrist-Hashing [109] today. Due to the vast amount of possible games (cf. Section 2.1.2), the high number of move options, the number of move choices in each single game and the complexity of estimating the value of arbitrary Go positions, the traditional brute-force search methods, like $\alpha\beta$ search, failed for Go. Accordingly, researchers focused on subproblems like Go for smaller board sizes [99] or the determination of the so-called *life-and-death* state of local groups (cf. Section 2.1.1) [11]. A great deal of work was done by Thomas Wolf on so-called

[6] The human player that lost against Zobrist's Go program was called Mr. Cowan. At the time of the experiment, he was an undergraduate at the University of Wisconsin with a total playing experience of 5 games. Zobrist's program was written in ALGOL for the Burroughs B5500 computer.

tsume-go, a class of mostly local Go problems about life-and-death, ko, capturing races and others. He developed the popular program *GoTools* [106] that is able to solve a number of difficult tsume-go problems. The improvements towards tackling local subproblems naturally lead to investigations on combinatorial game theory, that focuses on dividing games into subgames, solve them individually to afterwards draw conclusions about the original problem [77][56]. Another source of improvement was the study of abstractions of Go positions [44] and moves by shape patterns [12][97]. Both techniques are still used by modern Go programs. Due to the inapplicability of brute-force search methods, the most successful Go programs before 2006 were generally knowledge intensive systems composed of pattern based predictors, rule-based expert systems, local $\alpha\beta$ searches and databases of common move sequences, called *joseki*. Among the most popular Go programs of that time are *The Many Faces of Go* [38](MFoG), *Go++* and *Handtalk*. MFoG was developed by David Fotland and remains one of the strongest Go programs until today, having adopted to modern algorithms. In 2003, Erik van der Werf et al. published their results on solving 5×5 Go [102] with a search based approach implemented in their program *MIGOS* (MIni GO Solver). So far, 5×5 is the largest squared board size for that Go was solved.

Despite all of the before mentioned work, from the 1960s onwards, it was the case that researchers and developers of Go playing programs had major difficulties to approximate the strength of experienced human players. While for computer chess, the development of evaluation functions to be used with the $\alpha\beta$ search method and a number of enhancements for the $\alpha\beta$ search method itself, lead to programs that were able to defeat the world's strongest humans, for decades of research, the strength of computer Go programs stayed significantly below those of strong human amateur players. A vital breakthrough was achieved only in 2006, when Rémi Coulom managed to win an international Computer Go tournament with his Go playing program CrazyStone, implementing a non-traditional *Monte-Carlo Tree Search* (MCTS) approach.

2.3.2 Computer Go as Fertile Ground for Monte Carlo Tree Search

The development of plain Monte-Carlo (MC) approaches in Computer Go is said to have started in 1993, when Brügmann [20], a German physicist, presented in his seminal paper *Monte Carlo Go* a simulated annealing based method to search for good moves in his Go playing program *Gobble*. Already in 1987 and 1990, Abramson proposed [1] and analysed [2] an MC approach, the *expected-outcome* model, for the design of generic evaluation functions that can be used, e.g., with depth limited minimax search. It was only in 2003 that Bouzy and Helmstetter [17] took up Brügmann's approach and extended it with an efficient way to grow and store a game tree in memory. Don Dailey achieved a first success with his Monte-Carlo simulation based Go program botnoid, when placing 3rd out of 6 at the 1st KGS Computer Go tournament in April 2005. An advanced memory management was used by the before mentioned Rémi Coulom [27] who dominated the 2006 Computer Go Olympiade on the 9×9 game board with his Go program *CrazyStone*. Also in 2006, Kocsis and Szepesvári published a highly influential paper about the *Upper Confidence Bounds applied to trees* (UCT) algorithm [60] that extends the Upper Confidence Bounds (UCB) policy for the multi-armed bandit problem (MAB) proposed earlier by Auer et al. [6] to tree search. UCT in combination with Coulom's memory management was in turn used by Gelly et al. [43] in their Go program *MoGo* that became even stronger than CrazyStone at that time. The achievements of both these programs marked the breakthrough of MCTS in Computer Go and the success of MoGo made UCT the most popular MCTS algorithm. In the wake of the enormous success of MCTS in Computer Go numerous researchers adopted the UCT algorithm in subsequent years. In a recent survey of MCTS methods, Browne et al. [19] listed almost 250 MCTS related publications originating only from the last seven years, which demonstrates the popularity and importance of MCTS. MCTS is currently emerging as a powerful tree search algorithm yielding promising results in many search domains requiring only little or no domain knowledge at all. MCTS also performs remarkably well for games other than Go, such as connection games Hex [5] and Havannah [98], combinatorial games Breakthrough [67] and Amazons [66] as well as General Game Playing [37] and real-time games. Apart from games, MCTS finds applications in combinatorial optimization [60], constraint satisfaction [86], scheduling problems [78], sample-based planning [41] and procedural content generation [73].

CHAPTER 3

Monte-Carlo Tree Search

This chapter is devoted to the formal introduction of the family of Monte Carlo Tree Search (MCTS) algorithms. We will describe the basic framework in detail and briefly discuss a number of extensions that proved to be beneficial for Computer Go and other domains. An overview of the history and development of MCTS based algorithms can be found in Section 2.3.2.

3.1 Terminology

As induced by the name itself, MCTS is designed to search in domains that can be represented as trees. Throughout this thesis, we will mostly consider game trees as search domains (cf. Section 2.2.1 for the definitions of *tree* and *game tree*), or more strictly, game trees of combinatorial games. To motivate our choice and characterize the set of domains where MCTS is applicable, we extend our collection of formal definitions to *Markov Decision Processes* and *Combinatorial Games.*

Known from, e.g., decision theory, Markov decision processes (MDPs) are used to model sequences of decisions that lead to some kind of reward for the decision maker.

Definition 6 (Markov Decision Process) *A Markov Decision Process is a 4-tupel $MDP := (S, A, T, r)$ of a set of states S, a set of actions A, a transition model $T(s, a, s')$ that determines the probability of reaching state $s' \in S$ when* *choosing action $a \in A$ in state $s \in S$, and a reward function $r(s, s')$, that assigns some reward for transitions from state s to s'.*

In any state, a decision maker has to choose one action of the set of possible actions A that will lead him to a subsequent state, based on some probability distribution given by the transition model T. For any state transition, the decision maker obtains some reward. The uncertain transitions, based on a probability distribution, make MDPs so called discrete time stochastic control processes, that lead to a, possibly infinite, sequence of states and actions $(s_1, a_1, s_2), (s_2, a_2, s_3), \ldots, (s_{n-1}, a_{n-1}, s_n)$ and a corresponding total reward $R = \sum_{t=1}^{n-1} r(s_t, s_{t+1})$.

A general problem to solve for MDPs, is the identification of a *policy* for the decision maker, typically denoted by $\pi : S \to A$, that, for any given state, determines the corresponding action that must be selected in order to maximize the decision maker's expected total reward. MDPs are among the most general search domains that can be tackled with the MCTS approach (cf. [60]). We will now introduce a more specific class of problems that includes games like Chess and Go. It is the class of problems that were subject of intensive MCTS related research in recent years.

In game theory, a *Combinatorial Game* describes a decision process with two decision makers (i.e. players) involved that pursue opposed objectives:

Definition 7 (Combinatorial Game) *We write a two-player game as a tuple $G := (S, A, \Gamma, \delta, r)$, with S being the set of all possible states (i.e. game positions), A the set of actions (i.e. moves) that lead from one state to the next, a function $\Gamma : S \to \mathcal{P}(A)$ determining the subset of available actions at each state, the transition function $\delta : S \times A \to \{S, \emptyset\}$ specifying the follow-up state for each state-action pair where $\delta(s, a) = \emptyset$ iff $a \notin \Gamma(s)$ and a reward function $r : S_t \times P \to (-1, 1)$ assigning a reward to each terminal state $S_t := \{s \in S | \Gamma(s) = \emptyset\}$ for each of the two players $p_1, p_2 \in P$. If $r(s, p_1) = -r(s, p_2)$ for any $s \in S$, we call G a zero-sum game. We call G a Combinatorial Game if it is a zero-sum, two player game, where actions are applied in alternate turns by both players and the entire information about the game (i.e. S, A, Γ, δ and r) is available to both players at any time.*

Combinatorial games have a clear and simple definition, making them especially interesting as testbeds for artificial intelligence research. At the same time, they cover a large number of real-life games that are generally accepted to be highly complex. Given some start state $s_0 \in S$, a combinatorial game can be represented as a game-tree. We devote the next section to an introdution of the basic algorithmic framework of Monte-Carlo Tree Search and its application to combinatorial games.

3.2 Basic Algorithmic Framework

Monte-Carlo Tree Search is a simulation based search algorithm. Given an MDP and a current state $s_0 \in S$, a simulation is a sequence of states and actions starting at s_0, combined with the corresponding reward obtained. The state-action sequence is hereby generated by a decision maker using some partly randomized policy. Respectively, for a combinatorial game, a simulation is a path in the game tree starting at state s_0 denoted by the root node, to some terminal state with its associated reward. Again, the path is chosen by some partly random selection of actions in each state along the path. The central part of MCTS is the computation of such simulations.

Breaking with the initial randomness in the selection policy, MCTS starts to guide future simulations based on the outcomes of former ones. This imposes the need to store statistics about simulation rewards for distinct states. As typically a large number of different states are visited in the course of a search run, storing reward statistics for all visited states creates a huge demand for memory. Hence, for feasibility reasons, practical MCTS implementations only keep reward statistics for near-root tree nodes in memory. A way for an efficient memory usage was proposed in [27]. Here, a search tree representation T is generated in memory by initially starting with the root node only and expanding T by adding one node in each simulation step according to some selection policy π_t. This results in efficient and predictable memory usage, as the memory tree T likely grows in the most interesting branches and a maximum of one tree node is added with each additional simulation. Once a simulation leaves T, a randomized heuristic policy π_p is used for action selection until a terminal state is reached, i.e., a leaf of the game tree. We denote the randomized heuristic policy π_p as *playout policy* and the history dependent one used for nodes covered by T as *in-tree policy* π_t.

Essentially, MCTS algorithms break down into the four building blocks *Selection*, *Expansion*, *Playout* and *Update* as shown in Figure 3.1. These building blocks, that together make up a single simulation, are repeated in an endless loop. We can stop MCTS after any iteration to obtain the best action found so far, making MCTS a so called any-time algorithm. A search run is typically limited by some fixed number of simulations or a time budget.

Algorithm BasicMCTS on page 22 shows a pseudo-code representation of the basic MCTS algorithmic framework. The block around line 5 implements the selection phase. T is expanded with the first state reached, that is not already covered by

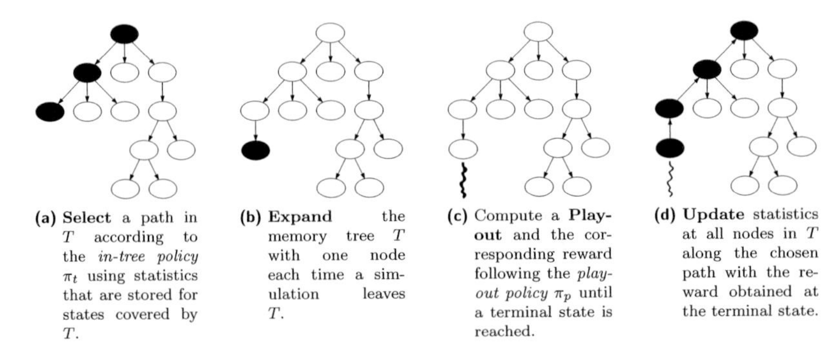

(a) **Select** a path in T according to the *in-tree policy* π_t using statistics that are stored for states covered by T.

(b) **Expand** the memory tree T with one node each time a simulation leaves T.

(c) Compute a **Playout** and the corresponding reward following the *playout policy* π_p until a terminal state is reached.

(d) **Update** statistics at all nodes in T along the chosen path with the reward obtained at the terminal state.

Figure 3.1: Building blocks of MCTS.

T in line 9. In line 10, the playout and its corresponding reward is computed. The reward is thereafter used to update the statistics of all states in T that were visited during the simulation, represented by the call of an update procedure in line 11.

Algorithm BasicMCTS: Basic MCTS-Algorithm.

Data: A combinatorial game $G := (S, A, \Gamma, \delta, r)$.

Input : A state $s_0 \in S$ and a time limit.
Output: An action $a \in \Gamma(s_0)$.

1 $T \leftarrow s_0$; // Initialize memory tree
2 $d \leftarrow 0$; // Initialize depth variable
3 **while** *Time available* **do**
4   **if** $s_d \in T$ **then**
5         $a_d \leftarrow \pi_t(s_d)$; // Action selection following in-tree policy
6         $s_{d+1} \leftarrow \delta(s_d, a_d)$;
7         $d \leftarrow d + 1$;
8   **else**
9         $T \leftarrow T \cup s_d$; // Expand memory representation of search tree
10   reward $\leftarrow$ Playout (s_d); // Compute a playout
11   Update (reward); // Update statistics for visited states and actions
12        $d \leftarrow 0$; // Start a new simulation
13   **end**
14 **end**
15 **return** $\text{argmax}_{a \in \Gamma(s_0)} N_{s_0,a}$; // Return action that was simulated most often

Looking at this general framework, we identify a number of points that need a more precise description when it comes to an actual implementation for a specific search domain. One central part, is the definition of the in-tree policy π_t for action selection based on former simulation rewards. A key task of the policy is the handling of the so called *exploitation-exploration dilemma*. The dilemma describes the opposed objectives of targeting at the exploitation of actions that

Function Playout(s)

Data: $G := (S, A, \Gamma, \delta, r)$ as described in the text.

Input : A state $s \in S$ that is the starting point of the playout.
Output: The reward obtained at the terminal state of the playout.

```
1 while Γ(s) ≠ ∅ do
2   | a ← π_p(s);                 // Action selection following playout policy
3   | s ← δ(s, a);
4 end
5 return r(s);                    // Return simulation reward at terminal state
```

Procedure Update(reward)

Data: The sequence of states and actions $(s_0, a_0), (s_1, a_1), \cdots, (s_d, a_d)$ visited by the simulation.

Input : A reward value.

```
1 i ← 0;
2 while i ≤ d do
3   | UpdateStateActionStatistics ((s_i, a_i),reward);
4   | i ← i + 1;
5 end
```

yielded good rewards in the past while aiming at exploring other actions because they might yield even better payoffs in the long run. As this basic problem is unrelated to a specific search domain, we will discuss it here, in the context of the basic MCTS framework.

3.2.1 Bandit Based In-Tree Policy π_{UCT}

An in-tree policy π_t is used to determine an action to be taken at a given state. Hence, it is of the form $\pi_t : S \rightarrow A$. To decide for an action, the policy can make use of the history of actions applied at the given state by former simulations as well as the history of the corresponding rewards obtained. Accordingly, it should be regarded as part of the policy design to determine the kind and extent of data about former simulations that is stored for each state.

We focus our discussion to the most prominent general MCTS in-tree policy that yielded great results for a large number of search domains. The policy was proposed for the use with MCTS in Kocsis and Szepesvári's seminal paper on bandit based Monte-Carlo planning [60] and is today known as the UCT policy. UCT is short for *Upper Confidence bounds applied to Trees*. The development of UCT is based on prior research on the multi-armed bandit problem (MAB), that basically formalizes the before mentioned exploitation-exploration dilemma.

Definition 8 (Multi-Armed Bandit (MAB)) *A K-armed bandit is a number of K independent gambling machines (i.e. single arm bandits) that each yield independent and identically distributed rewards from a fixed but unknown distribution with unknown expectations μ_i, for $1 \leq i \leq K$.*

The Multi-Armed Bandit problem is to determine a gambler's policy π, for subsequent play of one arm of the bandit at a time, with the objective of maximizing the gambler's reward. Hereby, the policy can use the information about rewards obtained with each arm over time. The notion of the *regret* is used to describe the loss of the gambler resulting from not always playing the optimal arm.

Definition 9 (Regret) *The regret of a policy π after n plays is given by*

$$\mu^* n - \mu_j \sum_{j=1}^{K} \mathbb{E}[T_j(n)] \quad \text{where} \quad \mu^* := \max_{1 \leq i \leq K} \mu_i,$$

$\mathbb{E}[\cdot]$ denotes expectation and $T_j(n)$ is the number of times arm j was chosen during the n plays.

In 1985, Lai and Robbins [61] proved that, for any optimal policy π that solves the MAB problem, a lower bound for the number of plays of any suboptimal arm j to be $T_j(n) \geq \ln(n) / \int p_j \ln(p_j/p^*)$. Here, p_j denotes the Probability Density Function (PDF) for the reward distribution of arm j, and p^* is the corresponding PDF of an optimal arm. The authors were even able to develop a policy that asymptotically attains this bound. However, the policy given by Lai an Robbins was rather hard to compute and the regret bound was only retained asymptotically. In 2002, Auer et al. [6] published efficiently computable and easy to implement policies with logarithmically bounded finit-time regret. Among those policies, the one they called UCB1, short for *Upper Confidence Bound 1*, became especially important for MCTS and will thus be presented here.

Definition 10 (UCB1) *The MAB policy μ_{UCB1} choses the next arm of a K-armed bandit after n former plays as follows:*

$$\mu_{\mathrm{UCB1}} := \underset{1 \leq j \leq K}{\operatorname{argmax}} \quad \bar{x}_j + \sqrt{\frac{2 \ln n}{T_j(n)}} \quad ,$$

where $\bar{x}_j$ denotes the average reward obtained with arm j during the former $T_j(n)$ plays. Initially, each arm is played once.

The square root term actually represents a one sided confidence interval for the unknown expected mean reward. Hence, the whole expression denotes a confidence bound for $\bar{x}_j$, giving rise to the policy's name. Looking at the formula with the exploitation-exploration dilemma in mind, we see the left hand term, i.e., $\bar{x}_j$, striving for exploiting arms that yielded high rewards in the past while the right hand term looks for exploring arms that might be underestimated by now. We see that, for each arm, the policy solely works on the average reward obtained and the number of times the arm was chosen in the past. Certainly, summing up the obtained rewards over time and counting the number of plays appears to be very efficient from the computational and memory usage perspective.

In 2006, Kocsis and Szepesváris [60] proposed the use of a variant of UCB1 for the MCTS in-tree policy μ_t, with the introduction of the before mentioned Upper Confidence Bounds applied to trees (UCT) algorithm. Thereby, each action selection step for states covered by the search space's memory representation $T \subseteq S$ is considered as an independent multi-armed bandit. Following our observation above, to apply UCB1, we need to maintain statistics about former simulation rewards in the form of the sum of obtained rewards and the number of times an action was chosen at each state covered by T. Accordingly, for each state $s \in T$ and its corresponding actions $a \in \Gamma(s)$ we maintain two variables in memory: $N_{s,a}$, that represents the number of simulations where action a was applied in state s and $W_{s,a}$ to keep the cumulative reward obtained by those simulations. We further write $N_s := \sum_{a\in\Gamma(s)} N_{s,a}$. Given this notation, the UCT policy is defined as follows:

Definition 11 (UCT) *The MCTS in-tree policy π_{UCT} selects the next action at any given state $s \in T$ by*

$$\pi_{\text{UCT}}(s) := \operatorname*{argmax}_{a\in\Gamma(s)} \quad \frac{W_{s,a}}{N_{s,a}} + C\sqrt{\frac{2\ln N_s}{N_{s,a}}} \quad ,$$

where $C > 0$ is a constant.

The novelty of this policy, compared to UCB1, is the introduction of the constant factor C for the exploration term. Due to the recursive use of μ_{UCT} in the search tree, the sequence of experienced rewards becomes non-stationary, i.e., the expected mean reward might drift over time. Kocsis and Szepesvári have proven that appropriate values for C can be determined at each state, that ensure a convergence of the sample means towards the correct expectations with overwhelming probability. Mind that C, being a factor of the right hand term of the policy equation, directly influences the weighting between exploitation and exploration, where larger values lead to more exploration.

We obtain the UCT algorithm from the basic MCTS algorithmic framework by replacing π_t with π_{UCT} in Algorithm BasicMCTS and by replacing the call to the update procedure for state-action pairs in procedure Update (cf. page 23) at line 3 by a call to UpdateStateActionStatisticsUCT, that is given below.

Procedure UpdateStateActionStatisticsUCT((s,a),reward)

input : A state-action pair and a reward value

1 $N_{s,a} \leftarrow N_{s,a} + 1$;
2 $W_{s,a} \leftarrow W_{s,a} +$ reward;

At each search tree node, the UCT policy targets at maximizing the cumulative reward observed. Accordingly, for two player games like Go, the reward must be appropriately inverted during the statistic's update. In case of Go we use a reward function that returns 1 for each terminal position in that Black has won and 0 for a win of White. In case of a draw, 0 or 1 is returned uniformly at random leading to an expected sample mean of 0.5. Hence, during the statistic's update phase, an obtained reward r is added to W_{s_i,a_i} for all states s_i in that the black player is to move, while $1-r$ must be added at all other states for the UCT policy to work as intended. In the remainder of this thesis we will concentrate solely on this special case of binary rewards, if not stated otherwise.

3.2.2 First Play Policy

A problem that arises when using UCT as introduced above, is the need for statistics for all actions available at a given state to select an action. Neither $\bar{x}$ nor the confidence bound is computable for actions when they appear as possible choices for the first time. Following the definition of UCB1 (cf. Definition 10 at page 24), prior to applying the actual selection formula given, each arm, i.e., each action, must be played once. For example for the game of Go, with an average number of about 200 move choices available in each position, the overhead for such initializations becomes high and requires to search a large number of potentially bad variations. Consequently, the selection policy at recently discovered states turned out to be crucial for the performance of UCT.

A number of different approaches for tackling this issue were proposed. One of the most straight forward approaches, proposed in [43], is the assignment of a fixed, so called *first play urgency* to each not yet applied action available at a state, that is used as the action's value in π_{UCT}. An alternative is the assignment of prior values to $N_{s,a}$ and $W_{s,a}$. If expert knowledge is available that allow for an estimation of the relative value of actions available at a state, the assignment of adequately derived prior values can lead to substantial improved search quality. While the fraction $W_{s,a}/N_{s,a}$ must reflect the action's estimated value, the choice of $N_{s,a}$ defines the initial weight of the prior. If the prior's value for $N_{s,a}$ is initially chosen to be, e.g. 5, the quality of the action value's estimate is expected to be identical to the quality of the average reward obtained from 5 simulations. Chaslot et al. [25] proposed to add another term to the original UCT formula that yields an initial action value that is derived from expert knowledge and that is down-weighted with increasing numbers of samples. They called their approach Progressive Bias (cf. Section 3.3.2).

3.2.3 Handling of Transpositions

In some search domains, it might be possible to reach the same state through different state-action sequences. Hence, eventually the resulting search space is a directed acyclic graph instead of a tree. We call such multiple paths leading to the same state *transposition*, and a corresponding data structure that contains such transpositions a *transposition table.* Accordingly, the search space representation T, constructed during the MCTS search should be organized as a transposition table. Transposition tables can be efficiently implemented in form of hash maps. An obvious requirement is the feasibility of constructing a hash function for the search domain's state space, i.e., an injective function $h : S \rightarrow \mathbb{N}$, that uniquely maps each state $s \in S$ to a scalar value. Some authors of UCT based MCTS searchers prefer to store the statistics about an action a available in state s, i.e., $N_{s,a}$ and $W_{s,a}$, in the transposition table entry corresponding to the unique subsequent state $s' = \delta(s, a)$. In contrast, we propose to store the statistics of all possible action choices with the corresponding states in that the actions where applied. That is, we argue for storing $N_{s,a}$ and $W_{s,a}$ for all $a \in \Gamma(s)$ in the transposition table entry of state s, pointing to the fact that π_{UCT} will regularly iterate over all such values, leading to reduced overhead due to fewer transposition table accesses and more effective usage of memory caches.

In 1969, Albert L. Zobrist published a hashing method for the game of Go that, given the hash value of a Go position, allows for an efficient generation of all hash-values of the corresponding rotated, mirrored and/or color inverted positions only with computations on the hash-value itself. This can be helpful as, e.g., two Go positions that are rotations of each other will have the same appropriately rotated best action. The method is today known as Zobrist-Hashing [109]. In addition to that, in the MCTS context, Enzenberger and Müller [35] described a lock-free implementation of a transposition table based on the use of atomic instructions, that allows for very efficient concurrent access in shared memory environments. We deal with transposition tables and concurrent accesses to them in detail in Section 4.2.3.

3.3 Enhancements to Basic MCTS

As part of the MCTS revolution in Computer Go, several enhancements to the original algorithm were developed, generally in form of modifications to the in-tree policy and the kind of statistics its design is based on. Among the most general and powerful are variants of so called *All-Moves-As-First* heuristics, that allow for faster generation of low variance estimates for state-action values. In addition, some progressive pruning strategies were proposed and investigated.Those strategies initially limit the number of actions considered and progressively widen its scope to more actions with increasing number of simulations. These strategies need a heuristic ordering of the available actions that is also crucial for its efficiency. We will present some of the most prominent variants of such enhancements in the remainder of this section.

3.3.1 RAVE: Rapid Action Value Estimate

RAVE is short for Rapid Action Value Estimate and was proposed in 2007 by Gelly et al. [42]. It is a form of the before mentioned family of *All-Moves-As-First* (AMAF) heuristics. We will first briefly introduce and discuss the basic AMAF idea to later on derive the RAVE approach from basic AMAF.

AMAF was already mentioned by Brügmann [20] in 1993. It is based on the idea that, in the course of an MCTS simulation, at a given state s, not only the action applied in s leads to the simulation's final reward, but the whole sequence of actions that were applied until a terminal state was reached. Hence, if some of those actions were already applicable in s, their corresponding statistics can be updated in the same way as it is done for the actually chosen action in s. Or, saying it in the words used in the heuristics name: during the update process, all actions in the sequence of actions applied in the course of a simulation are treated as if they were applied first.

Our regular state-action statistics as being used for UCT, are made up by two counters, one that counts the number of simulations ($N_{s,a}$) that applied action a in state s, while the other ($W_{s,a}$) sums up the rewards obtained at the end of each such simulation. For AMAF, instead, we maintain two counters, $N_{s,a}^{\text{AMAF}}$ and $W_{s,a}^{\text{AMAF}}$ for each state-action that are updated not only for actions that are selected directly at a given state, but also for actions selected at any time in the remainder of a simulation. Thus, consider a simulation with the following sequence of state-actions: $(s_1, a_1) \rightarrow (s_2, a_2) \rightarrow (s_3, a_3) \rightarrow \cdots \rightarrow (s_k, a_k)$ that yields the reward $r(s_k)$. The sequence will end up in increments of the following counters: N_{s_i,a_j}, W_{s_i,a_j}, for all $1 \leq i \leq j \leq k$ with $a_j \in \Gamma(s_i)$, i.e., action a_j is available at state s_i and was applied during the simulation at state s_i or later. In case of two or more player games, only actions applied by the same player should be considered for AMAF updates.

AMAF can lead to much faster low variance estimations for state-action values as, for the same number of simulations, more reward samples might be incorporated compared to the standard UCT update procedure, depending on the search domain. However, it is important to note that this is achieved by sacrificing the entire sequentiality information during the reward update. In search domains where the exact sequence, in that actions are applied, is irrelevant, AMAF can greatly increase the search quality for fixed numbers of simulations. In the case of combinatorial games, however, the sequentiality is generally not negligible and ignoring it can lead to severe estimation errors. In case of Go, it appears that the sequentiality can be ignored only to a certain extend. Moves that appear reasonable at a certain state, might remain reasonable for longer, if not played immediately. In the end, however, the correct order of the moves remains extremely important.

The enhancement of RAVE to AMAF, is the idea of using a linear combination of both statistics, i.e.,

$$(1-\beta)\frac{W_{s,a}}{N_{s,a}} + \beta\frac{W_{s,a}^{\text{AMAF}}}{N_{s,a}^{\text{AMAF}}} \text{ with } \beta \in [0,1] ,$$

and to move the weighting of the statistics with increasing numbers of observations towards $W_{s,a}/N_{s,a}$. This makes sense, as in the initial phase, the regular UCT statistics have high uncertainties whereas the AMAF statistics become reliable more early. Over time, the UCT statistics that preserve the sequentiality information, become more valuable and consequently should supersede the AMAF values. Hence, in the formula above, β becomes a steadily decreasing function over the number of simulations (either $n_u := N_{s,a}$ or $n_a := N_{s,a}^{\text{AMAF}}$), i.e., $\beta : \mathbb{N} \to [0,1]$. The actual choice of β is key and was initially proposed in [42] as

$$\beta(n_u) := \sqrt{\frac{k}{3n_u + k}} ,$$

for some equivalence parameter k, that must be selected as the number of simulations that is assumed to make the quality of the regular UCT statistics equivalent to the AMAF statistics quality.

Later on, David Silver [91] presented another formula for β that lead to an increased search quality for several Go engines that used RAVE at that time:

$$\beta(n_u, n_a) := \frac{n_a}{n_u + n_a + 4n_u n_a b_a^2} .$$

It was derived be making some simplifying assumptions on the underlying reward distributions and the objective of minimizing the mean-squared error of the RAVE value under these assumptions. The formula depends on both, the number of updates included in the basic UCT statistics (i.e. $n_u = N_{s,a}$), as well as those included in the corresponding AMAF statistics (i.e. $n_a = N_{s,a}^{\text{AMAF}}$). There still remains a constant b_a that represents the AMAF bias, i.e., the estimated error between the AMAF and regular UCT value: $b_a = W_{s,a}^{\text{AMAF}}/N_{s,a}^{\text{AMAF}} - W_{s,a}/N_{s,a}$.

By adding the exploration term, we obtain the resulting RAVE policy:

Definition 12 (RAVE) *The RAVE policy π_{RAVE} selects the next action at any given state $s \in T$ by*

$$\pi_{\text{RAVE}}(s) := \underset{a \in \Gamma(s)}{\text{argmax}} \quad (1 - \beta(n_u, n_a))\frac{W_{s,a}}{N_{s,a}} + \beta(n_u, n_a)\frac{W_{s,a}^{\text{AMAF}}}{N_{s,a}^{\text{AMAF}}} + C\sqrt{\frac{2 \ln N_s}{N_{s,a}}} \quad ,$$

where $C > 0$ is a constant,$n_u = N_{s,a}$ and $n_a = N_{s,a}^{\text{AMAF}}$.

While we concentrate on RAVE throughout this thesis, investigations on further variants of AMAF in the context of Computer Go can be found, e.g., in [48].

3.3.2 PB: Progressive Bias

Major improvements in Computer Go were made by the use of expert knowledge gathered from large databases of Go games using machine learning. Once expert knowledge is available and can be used for probabilistic move prediction, it is crucial how to use this additional information in MCTS. The *Progressive Bias* (PB) [25] approach proposes the use of move ranking information to assign prior values to newly discovered tree nodes that are evened out when regular simulation statistics become more meaningful. Given some function $H : S \times A \to \mathbb{R}$, that assigns a heuristic value to each state-action pair (e.g. following some expert knowledge), they propose a slightly modified UCT selection policy of the following form:

$$\pi_{\text{PB}}(s) := \underset{a \in \Gamma(s)}{\text{argmax}} \quad \frac{W_{s,a}}{N_{s,a}} + C\sqrt{\frac{2 \ln N_s}{N_{s,a}}} + \frac{H(s,a)}{N_{s,a}} \quad .$$

Here, the last term in the sum is the progressive bias term. By using a function H that is derived from the playout policy π_p, a smooth transition between the in-tree part and the playout part can by realized. In our own experiments we obtained satisfying results with

$$H(s,a) = \frac{\pi_p(s,a)}{\sum_{a' \in \Gamma(s)} \pi_p(s,a')} \quad .$$

3.3.3 PW: Progressive Widening

Both, Chaslot et al. [25] and Coulom [28] independently proposed progressive strategies called *Progressive Unpruning* and *Progressive Widening* (PW) respectively, that both, for each state, initially starting with a small number of actions, progressively widen the view to more actions with increasing number of simulations. Like for PB, these strategies need a heuristic ranking of the available actions that is also crucial for its efficiency. We denote the ranking function with $R : S \times A \rightarrow \mathbb{N}$ and expect it to assign ranks to state-action pairs, that induce a strict order of all actions available at a state s. Hence, there exists an indexing of the set of all actions applicable at state s, with $R(s, a_1) > R(s, a_2) > \cdots > R(s, a_{|\Gamma(s)|})$.

During the in-tree phase of a standard UCT search, the idea is to prune all actions applicable at state s, except the highest ranked action a_1. With increasing numbers $N_s = \sum_{a \in \Gamma(s)} N_{s,a}$ of simulations that visit state s, the number of actions being considered for selection is progressively increased. While Chaslot et al. [25] give no details about the actual *unpruning* policy, Coulom [28] proposes to successively add action a_{n+1}, after t_{n+1} simulations have visited state s, where $t_{n+1} = t_n + 40 \cdot 1.4^{(n-1)}$ and $t_1 = 0$.

3.4 Playout Policies π_p

The playout policy (alos known as default policy or rollout policy) is a critical part of MCTS implementations. At the end of a playout, when a terminal state is reached, it yields the final simulation reward[1]. The simulation rewards are in turn the central data of MCTS algorithms, as future simulations are guided following computations on them. Especially for search domains with heavily delayed payoffs, i.e., for domains that require long sequences of action selections until a final reward is obtained, the design of the playout policy becomes a key challenge. This is because, the more actions need to be applied in the course of a simulation, the larger will be the ratio of actions applied by the playout policy to action applied by the in-tree policy. Consequently, a lot of research was carried out on playout design. While the policy development is often highly domain specific, again, we will focus our attention on research related to Computer Go and give an overview of the work published in recent years.

[1] This is the case for combinatorial games at least. For other search domains, like MDPs, rewards might be obtained with each state transition.

3.4.1 Random Policies

One of the most basic kinds of MCTS playout policies is the completely random one. It selects actions uniformly at random from the set of available actions at each state. Although such a policy can be described as kind of stupid, it has a number of features that are worth to mention:

- Random policies have mostly *negligible sequential dependencies* between consecutive action decisions. This makes them especially attractive for use with AMAF heuristics. This is because the sequence of action choices made throughout each single playout might be permuted to large parts, still leading to valid state-action sequences with almost equal probability of being chosen by the same policy. Hence, the AMAF bias, as mentioned in Section 3.3.1 becomes close to zero.
- They obviously have *no need for domain knowledge.*
- They are *balanced*, i.e., in two or more player games, they play equally good (or bad) for both players. Non-random or semi-randomized policies can contain strategical imbalances that lead to biased rewards. This would be the case if, e.g., in a two-player game, the policy prefer attacking moves but not the respective defending answer moves.

The properties listed above make random policies appealing for fundamental research on MCTS.

In terms of simulations computable per time unit, Lucasz Lew implemented a highly efficient MCTS based search engine called libEGO[2] for the game of Go with random playouts, that greatly serves as a base for fundamental research [64]. In his implementation, he allows for all legal moves for selection, except those that would fill eyes of size one (cf. the definition of eyes in Section 2.1.1). We will denote this exception the *1-eye rule* in the remainder. By preventing the placement of stones into eyes of size one, groups can become alive and consequently the number of empty intersections, and thereby the number of legal moves, decreases over time. This considerably reduces the average number of move decisions to be made in the course of a simulation.

Apart from the above mentioned properties, however, completely random policies show bad performance when being used in Computer Go engines. Adjusting the distributions used for action selection towards favouring good moves and penalizing bad ones can lead to greatly improved search quality. Here, the classification of good and bad moves will obviously be made heuristically.

[2]Lucasz Lew's libEGO is available online at: `https://github.com/lukaszlew/libego`

3.4.2 Handcrafted Policies

The first published playout policies for MCTS based Go programs were handcrafted [27][43], i.e., the designer himself used his own Go knowledge to decide for the next move to play in each board configuration. Typically, handcrafted policies are fast and easy to compute leading to many simulations that are computable per time unit.

Coulom [27] used a random playout policy with the 1-eye rule described at the end of Section 3.4.1, as the base of his playout policy design. But instead of selecting moves uniformly at random, he increased the weight of moves to intersections that represent last liberties of groups. Those moves will lead to a capture if the corresponding group is of the opponent color, or might prevent captures in case the group is of the player's color. His policy furthermore favors some moves that explicitly create 1-eyes and discourages the choice of some obviously bad moves.

Gelly et al. [43] extended the policy of Coulom with the notion of local answer moves. While Coulom focused on moves to intersections that are last liberties of any group, all over the board, Gelly et al. concentrated on playing on intersections close to the last move. They therefore created a set of 3x3 shape patterns that defined *interesting local moves* and tested for them on the 8-neighborhood of the move played last. The 8-neighborhood of an intersection is the set of orthogonally and diagonally directly neighboring intersections. A 3x3 shape pattern is defined around some empty intersection and explicitly specifies the configuration of the empty intersection's 8-neighborhood. If interesting local answer moves exist, the policy selects one of them uniformly at random. Otherwise, Coulom's policy mentioned above is used as the fall back policy. By including this notion of locality, Gelly et al. were able to greatly improve the playing strength of their Go playing program Mogo. This kind of policy is sometimes denoted as *sequence-like* to emphasize the effect of playing very local or *Moggy-style* to stress the combination of policies that were initially used in the two Go playing programs Mogo and CrazyStone.

3.4.3 Machine Learning Based Policy Design

Crafting playout policies by hand naturally limits the number of parameters that can be used in a meaningful way to define a probability distribution used for move selection. Even further, the hand tuning of parameters generally requires numerous and time intensive experiments to measure the effect of parameter changes on the search quality, i.e., the playing strength in case of games. This makes machine learning interesting for automated or semi-automated design of policies. A number of approaches were investigated in the past to develop and automatically train models for move decisions. This includes approaches of supervised learning [96][28][105], where move values are learned from records of Go games played by strong humans as well as reinforcement learning techniques [16][42][92][93][51].

Among the reinforcement learning techniques, an especially interesting approach is the so called technique of *Simulation Balancing* (SB), initially described by Silver [93] and further investigated by Huang et al. [51]. Instead of learning a system that is able to play *good* moves in the game of Go, it focuses on learning a playout policy that leads to an average reward equal to a former defined target reward by stochastic gradient descent optimization. The approach is especially relevant as further investigations revealed that stronger playout policies, i.e., policies that select moves like strong Go players might do, can lead to decreasing search quality [16][23]. SB tries to circumvent this problem and lead to promising results for the 9×9 Go board but unfortunately not for larger board sizes [93]. A strong Go program implementing SB is the program Erica, by Shih-Chieh (a.k.a. Aja) Huang.

A number of supervised learning algorithms are discussed in more detail in Chapter 5.

CHAPTER 4

Parallel Monte-Carlo Tree Search[1]

In this chapter, we present and empirically analyze a novel data-driven parallelization approach for Monte Carlo Tree Search algorithms, targeting large HPC clusters with fast network interconnect. The power of MCTS strongly depends on the number of simulations computed per time unit and the amount of memory available to store data gathered during simulation. High-performance computing systems such as large compute clusters provide vast computation and memory resources and thus seem to be natural targets for running MCTS.

MCTS may be classified as a sequential best-first search algorithm [91], where *sequential* indicates that simulations are not independent of each other, as is often the case with Monte-Carlo algorithms. Instead, statistics about past simulation results are used to guide future simulations along the search space's most promising paths in a best-first manner. Looking for parallelization, this dependency could partly be ignored with the aim of increasing the number of simulations for the price of eventually exploring less important parts of the search space. Parallelization of MCTS for distributed memory environments is a highly challenging task since typically we want to adhere to simulation dependencies and, additionally, we need to store and share simulation statistics among computational entities.

Parallelization of traditional $\alpha\beta$ search is a pretty well solved problem, e.g., see [31][32][50]. While for $\alpha\beta$ search it is sufficient to map the actual move stack to memory, MCTS requires us to keep a steadily growing search tree representation in memory. Targeting at optimal utilization of the available resources, our algo-

[1] We presented parts of the work presented in this chapter before at different stages of its development: At the MCTS Workshop of the Int. Conf. on Automated Planning and Scheduling (ICAPS) in 2011 [45], as well as at the Euro-Par Conference 2011 [87]. A further article is currently under review at the IEEE Transactions on Computational Intelligence and AI in Games (T-CIAIG).

rithm spreads a single search tree representation among all compute nodes (CNs) and guides simulations across CN boundaries using message passing. In the context of MCTS, sharing a search tree as the central data structure in a distributed memory environment is rather involved and only few approaches have been investigated so far [13][107]. A comparable approach used with traditional $\alpha\beta$ search was termed transposition table driven work scheduling (TDS) [85]. Computing more simulations in parallel than distinct compute resources are available, allows us to overlap communication times with additional simulations.

We integrate our parallel MCTS approach termed Distributed-Tree-Parallelization in our state-of-the-art Go engine Gomorra [45] and measure its strengths and limitations in a real-world setting. Our extensive experiments show that we can scale up to 128 compute nodes and 2048 cores and, furthermore, give promising directions for additional improvement. The generality of our parallelization approach advocates its use to significantly improve the search quality of a huge number of current MCTS applications. We present an efficient parallel transposition table highly optimized for MCTS applications in time critical use cases. Even further, we propose the use of dedicated compute nodes (CNs) to support necessary broadcast operations, that allow for greatly reducing network traffic by multi-stage message merging.

All experiments are run on a homogeneous compute cluster with 4xQDR Infiniband interconnect. Our implementation is based on the OpenMPI library and exploits capabilities for low latency RDMA communication of tiny messages.

Our Go engine Gomorra has proven its strength at the Computer Olympiad 2010 in Kanazawa, Japan, the Computer Olympiad 2011 in Tilburg and several other international Computer Go tournaments. Gomorra is regularly placed among the strongest 6 programs and recently won a silver medal at the Computer Olympiad 2013 in Yokohama, Japan.

4.1 Related Work

In this section, we review parallelization approaches for MCTS that were developed and investigated in recent years. They can roughly but not strictly be grouped into approaches targeting shared memory systems, distributed memory systems and accelerator hardware like general purpose graphic processing units (GPGPUs) and field programmable gate arrays (FPGAs).

Depending on the target platform we distinguish different compute entities that can be single cores of a multi core processor or entire nodes of a compute cluster that might contain a number of multi core processors each. In the remainder of this section, we will use the term compute entity (CE) for simplicity when we write about distinct computing units that might either be single cores or entire compute nodes.

4.1.1 Parallelization for Shared and Distributed Memory

The most prominent parallelization methods presented so far were termed Tree-Parallelization, Leaf-Parallelization and Root-Parallelization by Chaslot et al. [24] in 2008. Those terms are widely used until today, although different terms describing variations of these principle approaches can be found in the literature. Illustrations of these methods are depicted in Figure 4.1. In this figure, different colors represent simulations performed by different compute entities (CEs).

Tree-Parallelization

The Tree-Parallelization approach (cf. Figure 4.1a) is designed to be used on shared memory systems, that allow for sharing one search tree representation among several compute cores in shared memory. Each core performs one simulation at a time and updates a single shared transposition table that represents the search tree. A crucial part in practical implementation is the serialization of write accesses to the transposition table. This includes the adding of search tree nodes to the table in the expansion phase of MCTS, as well as the update of state-action statistics. Both forms of write accesses can be realized in a lock-free manner on modern shared memory systems by the use of atomic instructions. A detailed description of lock-free transposition tables is given in Section 4.2.3. Tree-Parallelization was first proposed and investigated in [24]. Later on, in 2009, Enzenberger et al. [35] published their work on a lock-free implementation of a transposition table for shared memory parallel MCTS in the context of Computer Go. Already in March 2008, Coulom posted his results and implementation details on a lock-free transposition table to the Computer Go mailing list[2]. Moreover, chess programmers have been using lock-free implementations of transposition tables for many years, cf. [55].

Leaf-Parallelization

Leaf-Parallelization (cf. Figure 4.1b) concentrates on the playout phase of MCTS that starts on the leafs of the search tree representation and exploits the independence of playout computations. While the in-tree part of MCTS makes use of

[2]Coulom's mail to the Computer Go mailing list about a lock-free transposition table in the MCTS context was sent in 2008 on the 21st of March and is available online at: `http://www.mail-archive.com/computer-go@computer-go.org/msg07611.html`

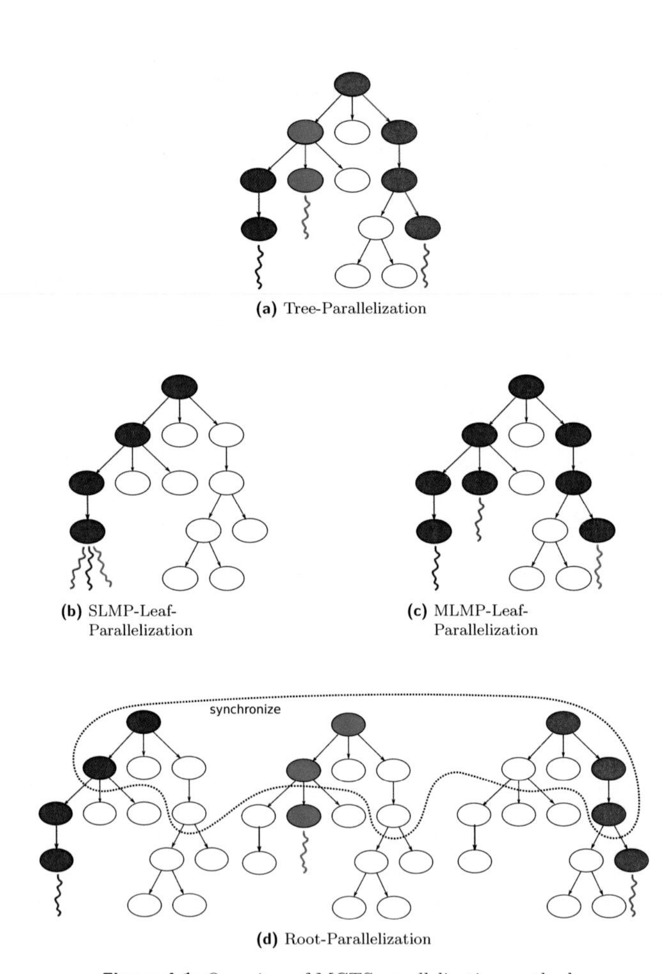

(a) Tree-Parallelization

(b) SLMP-Leaf-Parallelization

(c) MLMP-Leaf-Parallelization

(d) Root-Parallelization

Figure 4.1: Overview of MCTS parallelization methods.

former simulations' outcomes and thereby depends on them, the playout computation has no such dependencies. Furthermore, the computation of playouts that might include several hundred move decisions and a terminal reward computation, is generally assumed to be time consuming. Their independence, together with their computational overhead makes playouts especially attractive subjects for parallelization. We distinguish two types of Leaf-Parallelization:

Single Leaf Multiple Playouts (SLMP) was first proposed in [21] and is based on computing the in-tree part of the MCTS search on a single master computation entity only. After the expansion phase however all computation entities will independently compute a playout from the search tree node just expanded. The single master computation entity afterwards collects all playout rewards to continue with the update phase.

Multiple Leafs Multiple Playouts (MLMP) again divides the available computation entities into a single master entity and slave entities. As for SLMP, the in-tree part is solely handled by the master. But instead of having all slaves computing a playout from a single search tree node, the master delegates the playout computation to a single slave only, and asynchronously starts another simulation at the root of the tree until its expansion step is completed. Then again, the playout computation is delegated to another slave. MLMP was first proposed in 2008 by Cazenave et al. [22]. In 2010, Kato et al. [57] published further experiments with MLMP calling their implementation client-server style parallel MCTS.

The independence of the playout computations with one another makes Leaf-Parallelization attractive for distributed memory systems. In fact, all the data that a slave needs to perform a playout computation is a game position and the information about the player to move. On the other hand, the entire information a slave might return to the master is the sequence of actions chosen, the terminal reward and eventually the terminal position. Hence, at least for the game of Go, the size of data is small and can be efficiently communicated by message passing parallelization.

Root-Parallelization

While in all variants of Leaf-Parallelization, essentially a single MCTS search takes place, Root-Parallelization is based on multiple, mostly independent MCTS searches that are carried out in parallel. As depicted in Figure 4.1d, each CE grows its own search tree representation in memory by conducting an almost independent MCTS search. The searches are only *almost* independent, as the statistics at near-root nodes are synchronized from time to time. The concrete set of tree nodes that get synchronized as well as the particular frequency of synchronization are tunable parameters of the parallelization. Depending on the choice of the parameter values, different names were given to the resulting actual parallelization, among them Fast-Tree-Parallelization and Slow-Tree-Parallelization. Root-Parallelization was first proposed and investigated by Cazenave et al. [21] where only statistics about moves available at the root node were synchronized. Further variations were developed and investigated, e.g., in [24][13] and [9]. Like Leaf-Parallelization, Root-Parallelization is well suited and actually developed for message passing parallelization on distributed memory machines.

Among the above mentioned methods, Root-Parallelization is currently excelling for distributed memory systems [9]. One drawback of this method is that no effort is made at all to exploit the increased amount of memory available within a cluster. Instead, all CNs try to keep a nearly identical copy of the search tree representation. However, simply distributing the search tree across all cluster nodes would result in very costly remote read/write operations, slowing down the simulation dramatically [13].

In 2011, simultaneously and independent to our paper [45], Yoshizoe et al. [107] published an MCTS parallelization that manages the sharing of a single search tree representation on a distributed memory system. Both of us proposed a solution based on transposition table driven work scheduling (TDS) [85]. In contrast to us, Yoshizoe et al. developed a depth-first version of the UCT policy (df-UCT) that reduces the update frequency of statistics associated to often visited tree nodes. This is achieved by delaying and merging necessary backpropagations of simulation results. They restrict their scalability experiments to simulation rates achievable with artificial game trees and mention experiments regarding search quality improvements with real games like Go as an important direction for future work.

4.1.2 Parallelization for Accelerator Hardware

Focusing on more fine-grained parallelism, some attempts were made to move the playout computation to accelerator hardware, in particular FPGAs and GPGPUs. Concerning FPGAs we want to mention the work of Gao et al. [40] who presented a 167 staged pipeline architecture that implements two consecutive completely random move operations on a 9×9 Go board, one by the black and the other by the white player. This way, they were able to run 167 playouts in parallel and achieved a remarkable number of 1.56M playouts per seconds on a Cyclone III (EP3C120) FPGA from Altera running at 125 MHz. This is a speedup of about 15 over the current fastest CPU single core implementation of Łukasz Lew's Go engine libEGO that was reported to achieve about 100,000 playouts/second. Gao et al. used the playouts as an evaluation function with a traditional $\alpha\beta$ search that was also implemented on the same FPGA device. As their playouts are completely random, they only achieved a rather moderate overall playing strength.

In 2011, Rocki et al. [83][84] presented their Othello MCTS searcher for an NVIDIA TESLA C2050 GPU. They proposed a hybridized Root-Parallelization and SLMP Leaf-Parallelization. The GPU organizes a number of arithmetic logical units (ALUs) in so called multiprocessors (MPs) that can compute a number of SIMD threads in parallel. Following this special architecture, the authors propose to perform one MCTS search on each of the MPs but to use the SIMD threads computable at each MP to implement SLMP Leaf-Parallelization. Between the MPs they use Root-Parallelization to synchronize the results of all distinct MCTS searches. They call this GPU specific parallelization Block-Parallelization and claim to achieve a performance comparable to a hundred of CPU cores with a single GPU. Experiments were solely conducted in form of self-play experiments, making a comparison with state-of-the-art Othello programs difficult.

4.1.3 General Techniques for Scalability Improvements

During the years of research on parallelizing MCTS, some techniques were discovered that appear to generally lead to better scalability, almost independently of the actual form of parallelization used. We will present two of the most popular techniques in the following.

Virtual Losses

In [24], Chaslot et al. first observed substantial improvements in scalability by a quite simple modification to the original UCT algorithm. While the original algorithm exhibits a proper update phase in which all updates to tree node statistics are performed, the idea of *virtual losses* is to update the simulation count of tree nodes visited during a simulation already in the selection phase. The update phase then solely handles the increment of the visited nodes' reward accumulator. Recall, that for the UCT in-tree policy π_{UCT}, we compute the average reward observed from simulating some state-action by dividing the total sum of rewards by the number of simulations. Also, we defined the observed single simulation rewards for each state-action pair to equal 1 for a win of the player to move and 0 for a win of its opponent. Hence, an increment of the simulation count only produces statistics equal to a simulation where the player to move loses. Chaslot et al. called this effect a *virtual loss*, because, in fact, the statistics' update relating to the reward is only delayed, so that the virtual loss can still be updated to a real win. The concept of virtual losses encounters the problem of having a potentially large number of parallel simulations being guided along the very same path of the search tree by π_{UCT} only because of delayed statistic updates. This is achieved by slightly and temporarily penalizing moves for having been selected during an ongoing parallel simulation.

Virtual Wins

Baudiš and Gailly [9] observed further scalability improvements in their implementation of Root-Parallelization by assigning slightly different priors to the statistics of tree nodes on different compute nodes. They hereby artificially diversify the simulation guidance in the initial phase. The term *virtual wins* stems from the fact that one chooses the priors as if several simulations had been computed already that all led to a win.

4.2 Proposed Parallelization: Distributed-Tree-Parallelization

In this section, we present the central part of this thesis, our MCTS parallelization for distributed memory systems. It combines several techniques from the related work presented in Section 4.1, and furthermore, leverages concepts of data driven parallelization for efficient sharing of a transposition table in distributed memory systems [85][58]. Our extensive experiments, that are documented in Section 4.3, show that our parallelization can scale up to 128 compute nodes and 2048 cores, making it one of the best scaling parallelizations to date.

With our approach, we target modern HPC compute clusters with fast network interconnect, where each compute node (CN) contains a number of many-core CPUs. Looking for the best possible performance, we developed a hybrid parallelization that is capable of exploiting shared memory based low latency communication on the compute nodes while resorting to message passing for inter node communication. The use of compute clusters typically comes with large amounts of distributed memory. Efficiently exploiting not only the vast compute resources but also the entire memory available for improving the search quality was another central objective of our work.

Looking at the experimental analysis of an MCTS parallelization's scalability, we have to decide for a good way of measuring scalability. We are convinced, the most solid measure is the search quality improvement achieved with increasing compute and memory resources. By an implementation of our parallelization in a computer Go engine, we are able to quantify search quality with playing strength. Looking at the number of simulations computable per time unit (simulation rate) as well, we observe a non linear dependency of the simulation-rate to the search quality

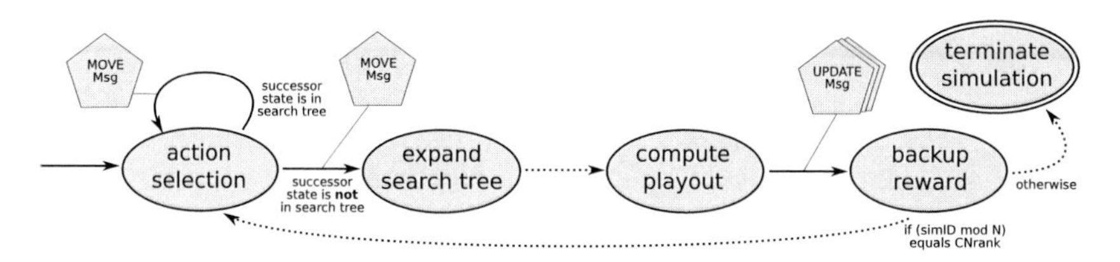

Figure 4.2: Finite state machine for distributed simulations (Worker FSM).

development. This supports our decision to focus on search quality rather than restrict measurements to simulation rates, as it is sometimes seen in literature.

At a high level, our parallelization is an approach to effectively realize Tree-Parallelization for distributed memory systems, while so long, Tree-Parallelization was applicable to shared memory systems only. This imposes the need for realizing a single search tree representation in distributed memory and developing an appropriate architecture that allows for low access times to this representation. We detail our parallelization and its implementation on a real system in the subsequent sections.

4.2.1 Distributed Simulations

The key technique of our approach is based on the transposition table driven scheduling (TDS) that was used with, e.g., $\alpha\beta$-search, or more precisely its variant MTD(f) before [58]. We spread a single search tree representation among the local memories of all CNs, moving computational tasks to the CNs that own the required data. Therefore, we break simulations into work packages that can be computed on different cores that not even need to be located on the same CN. Message passing is used to guide simulations over CN boundaries, i.e., the only way of communication used is by distinct messages that are explicitly sent to and received by remote CNs. We overlap necessary communication with concurrently computing more simulations than there are actual compute cores. This is a standard technique that is generally known as *latency hiding*.

Figure 4.2 illustrates our distributed simulation process as a finite state machine. The states represent work packages that make up the computational load. The work packages were derived from the building blocks of MCTS simulations as depicted in Figure 3.1. Dotted arrows in Figure 4.2 represent state transitions that always happen at a single CN. Solid arrows indicate a possible movement to other CNs and are annotated with the corresponding messages that have to be sent. During the in-tree part of a simulation, several action selection steps take place. Each of those steps, i.e., each application of π_{UCT}, can be computed without the need to communicate with other CNs by storing the statistics about all actions available in one state together in memory[3]. Between two consecutive action selection steps, a simulation may move to another CN through a MOVE message. The UPDATE message is sent to all CNs visited in the course of the simulation as statistics need to be updated on all of them.

4.2.2 Tree Node Duplication and Synchronization

An obvious bottleneck when implementing distributed simulations is a massive contention at memory locations that hold search tree statistics of near root nodes. This is because nodes near the root of the tree are naturally more likely visited by simulations compared to tree nodes at deeper levels. To address this imbalance, we duplicate the representation of such frequently visited tree nodes on all CNs as it is also done in the Root-Parallelization approaches. Duplicating and occasionally synchronizing frequently visited tree nodes on all CNs greatly reduces the average communication overhead of a single simulation and thereby speeds up the overall average time required for its computation. Furthermore, it prevents high congestion of single CNs that are visited by an above average number of simulations.

The duplication of frequently visited near-root nodes entails the need for regular synchronization of statistics stored with duplicated nodes. We therefore maintain additional counters $N^{\Delta}_{s,a}$ and $W^{\Delta}_{s,a}$ next to the $N_{s,a}$ and $W_{s,a}$ that were introduced in Algorithm BasicMCTS. $N^{\Delta}_{s,a}$ and $W^{\Delta}_{s,a}$ are used for accounting simulation visits and rewards that still have to be communicated to remote MPI ranks. After communicating them, the Δ-values are incorporated into the corresponding non-Δ variables. Hence, at any time the actual real visit count for a state-action pair (s, a) is obtained by $N^{\Delta}_{s,a} + N_{s,a}$. As we assume that all variables corresponding to children of the same tree node are stored together in memory, all those variables are synchronized at the same time with a single MPI message in order to minimize communication overhead.

We introduce the following four parameters to control the duplication and synchronization of frequently visited search tree nodes:

- N_{dup}: The minimum number of simulations that must have passed a tree node before it is eligible to be shared (i.e. duplicated).
- α: A factor to determine $N_{\text{sync}}(s, a) = \alpha(N^{\Delta}_{s,a} + N_{s,a})$ as the minimum value required for at least one $N^{\Delta}_{s,a}$ to start the synchronization process for node s.

[3]Storing statistics about all actions available at one state together at subsequent memory locations is in general an effective approach, as typical caching strategies assume linear memory access patterns. And in fact, the UCT policy π_{UCT} linearly accesses statistics of all actions available at a state.

- $N_{\text{sync}}^{\min}$: A lower bound for N_{sync}.
- $N_{\text{sync}}^{\max}$: An upper bound for N_{sync}. Hence, we actually have $N_{\text{sync}}(s,a) = \min(N_{\text{sync}}^{\max}, \max(N_{\text{sync}}^{\min}, \alpha(N_{s,a}^{\Delta} + N_{s,a})))$.

Parameter α can be selected with the objective to ensure a uniform reduction of the standard error of the tree node's mean reward per synchronization and to allow for less frequent synchronization of more settled node statistics. Let us consider a tuple $(s,a) \in S \times A$ of a node and an action that can be taken in node s with its corresponding values of the number of simulations $N_{s,a}$ that applied action a in node s and the number of those simulations $W_{s,a} \leq N_{s,a}$ that lead to a win. We can then compute the mean reward for simulations starting with the transition (s,a) and the corresponding standard error by Equation 4.1 and Equation 4.2 respectively.

$$\mu(s,a) = \frac{W_{s,a}}{N_{s,a}} \tag{4.1}$$

$$SE_{\mu} = \frac{\sigma}{\sqrt{N_{s,a}}}\,, \tag{4.2}$$

with σ denoting the standard deviation. Let us denote the standard error of the sample mean μ of a given state action pair (s,a) at the time of two consecutive syncronization points t and $t+1$ by SE_{μ}^{t} and SE_{μ}^{t+1} respectively. If we assume σ to remain unchanged and write $N_{s,a}^{\Delta}$ for the number of simulations performed between time t and $t+1$, we obtain the followig formulas for SE_{μ}^{t} and SE_{μ}^{t+1}:

$$SE_{\mu}^{t} = \frac{\sigma}{\sqrt{N_{s,a}^{t}}}\,, \tag{4.3}$$

$$SE_{\mu}^{t+1} = \frac{\sigma}{\sqrt{N_{s,a}^{t+1}}} = \frac{\sigma}{\sqrt{N_{s,a}^{t} + N_{s,a}^{\Delta}}} \tag{4.4}$$

We can then select a desired rate $p \in [0,1]$ of reduction of SE_{μ} for each update that yields $SE_{\mu}^{t+1} \leftarrow p \cdot SE_{\mu}^{t}$. Then, the following implications yield the maximum amount of simulations $N_{s,a}^{\Delta}$ that can be performed without reducing SE_{μ}^{t+1} to a value lower than $p \cdot SE_{\mu}^{t}$:

$$
\begin{aligned}
& & p \cdot SE_{\mu}^{t} &\le SE_{\mu}^{t+1} \\
&\Rightarrow & \frac{p}{\sqrt{N_{s,a}^{t}}} &\le \frac{1}{\sqrt{N_{s,a}^{t} + N_{s,a}^{\Delta}}} \\
&\Rightarrow & \frac{p^2}{N_{s,a}^{t}} &\le \frac{1}{N_{s,a}^{t} + N_{s,a}^{\Delta}} \\
&\Rightarrow & \frac{N_{s,a}^{t} + N_{s,a}^{\Delta}}{N_{s,a}^{t}} &\le \frac{1}{p^2} \\
&\Rightarrow & N_{s,a}^{\Delta} &\le \frac{N_{s,a}^{t}}{p^2} - N_{s,a}^{t} \\
&\Rightarrow & N_{s,a}^{\Delta} &\le (\frac{1}{p^2} - 1) N_{s,a}^{t} \\
&\Rightarrow & N_{s,a}^{\Delta} &\le \alpha N_{s,a}^{t} \qquad (4.5)
\end{aligned}
$$

Hence, given a desired reduction rate $p \in [0,1]$ we have $\alpha = (1/p^2) - 1$ and synchronization of node s will be triggered as soon as $N_{s,a}^{\Delta} \ge N_{\text{sync}}(s,a)$ for some $a \in \Gamma(s)$.

4.2.3 Distributed Transposition Table

During computation, workers need access to MC-simulation statistics of the tree nodes. As mentioned in Section 3.2.3, it is possible for some search domains that equal states are reached through different paths motivating the use of transposition tables. As transposition table we use a hash table H that is capable of storing up to |H| nodes of the search tree representation. We assume the existence of a hash-function $h : S \to \mathbb{N}$ assigning a unique and equally distributed hash value to each single state in the search space. One way for computing an index $I_{\text{H}}(s)$ into H for a search tree node $s \in S$ is given by:

$$I_{\text{H}}(s) = h(s) \bmod |\text{H}|$$

In a manner similar to [36], we distribute H among M MPI ranks on a cluster by storing for each rank $i \in \{0, \ldots, M-1\}$ a partial transposition table H_i with $|\text{H}|/M$ entries. An index to the distributed table is a tuple of a rank i and a local index I_{H_i} to H_i. Both are computed as follows:

$$i(s) = (h(s)/(|\text{H}|/M)) \bmod M \qquad (4.6)$$

$$I_{\text{H}_i}(s) = h(s) \bmod (|\text{H}|/M) \qquad (4.7)$$

Typical time settings for actual game play limit the time for each player to make all of his moves in a game. Among the obstacles towards actual game play with such time limits is the efficient deletion of parts of the transposition table elements, once a move was made and a new search run is to be started. In such cases, only a single subtree below a direct root-child, the one corresponding to the move that was actually made, must remain in the transposition table, while the other elements are likely not needed any longer. An active deletion or marking of such unneeded elements could easily end up in a lot of communication and might require a substantial amount of time. Saving time for this task allows for using more time with subsequent search runs.

We developed a method for efficient table lookup and insertion that does not require an explicit element deletion phase between subsequent search runs and efficiently implements an inherent locking mechanism for fast replacement of *old* elements. Therefore, we use a time-stamp with every table element to track the element's last access times and override the least recently updated element among the K upcoming indices in a linear probing manner, in case no free table entry was found.

Procedure TableLookup on page 48 shows a pseudocode representation of the table lookup operation that is used to find the actual table index, given a hash value. The procedure consists of two parts: the first part (line 1–22) looks for the best fitting index and the second part (line 23–43) updates the element's data and ensures that each element is only associated to a single hash value. In case no matching element was found in the first part, an insert takes place at the first free or the least recently updated table index, in case no free one is found. In order to be overridden, an old element must not have been accessed during the current search run. This is ensured by line 10 and in turn guarantees that no data of elements can be deleted that might be accessed by some thread in the current search run. In case we override an old element, we may need exclusive access to this element in order to clear/delete the element's data until it provides valid information again. We declare the element's time-stamp value 0 to be a special lock-value. Line 31 shows, how to use an atomic compare-and-swap operation (atomicCAS) with the lock-value to acquire exclusive access to a table element. The lock is released in line 34 implicitly, when the current value of the lookup counter, that is always greater than 1, is stored in the element's time-stamp variable. Note that in practical implementations, depending on the actual compiler and processor used, it might be necessary to add a compiler and/or CPU store memory barrier (also known as fence) right before line 34 to ensure all data was written before the implicit lock is released.

Procedure TableLookup

Data: An array T with size(T) elements that makes up the transposition table. Each element $T[i]$ is a 2-tuple (h, a) consisting of a hash value h and a volatile integer a tracking the elements last access time. For all i, $T[i].a$ is initialized with 1, indicating an empty table element. A global integer variable lookupCnt is incremented at each table lookup and thereby serves as a clock. startTime holds the value of lookupCnt at the time the search run started.

input : A hash value h

output: A table-index i^* and a return state $r \in \{\text{FOUND}, \text{FULL}, \text{BUSY}\}$

```
 1 i* ← 0;
 2 i_tmp ← 0;
 3 a_tmp ← MAXINT;
 4 i ← h mod size(T);
 5 for 10 times do
 6 |   i ← max(i, 1);                                   // make sure index 0 remains unused
 7 |   if T[i].a = 1 and a_tmp > 0 then
 8 |   |   i_tmp ← i;
 9 |   |   a_tmp ← 0;
10 |   else if T[i].a < startTime and T[i].a < a_tmp then          // found a free element
11 |   |   i_tmp ← i;
12 |   |   a_tmp ← T[i].a;
13 |   end
14 |   if T[i].h = h and T[i].a > 1 then                      // found an existing element
15 |   |   i* ← i;
16 |   |   break;
17 |   end
18 |   i ← (i + 1) mod size(T);
19 end
20 i ← T*;
21 if i = 0 then i ← i_tmp;
22 if i = 0 then return FULL;
23 atomicIncrement(lookupCnt);
24 a ← T[i].a;
25 while a ≤ startTime do
26 |   if a = 0 then
27 |   |   a ← T[i].a; continue;
28 |   end
29 |   if i* ≠ 0 then
30 |   |   atomicCAS (T[i].a, lookupCnt, a);
31 |   else if atomicCAS (T[i].a, 0, a) then               // atomicCAS(mem,newval,oldval)
32 |   |   clear(T[i]);
33 |   |   T[i].h ← h;                                   // delete/clear old elements data
34 |   |   T[i].a ← lookupCnt;                                             // release lock
35 |   end
36 |   a ← T[i].a;
37 end
38 if T[i].a > startTime and T[i].h = h then
39 |   T[i].a ← lookupCnt;
40 |   T_i ← i;
41 |   return FOUND;
42 end
43 return BUSY;                              // another thread updated our element T[i]
```

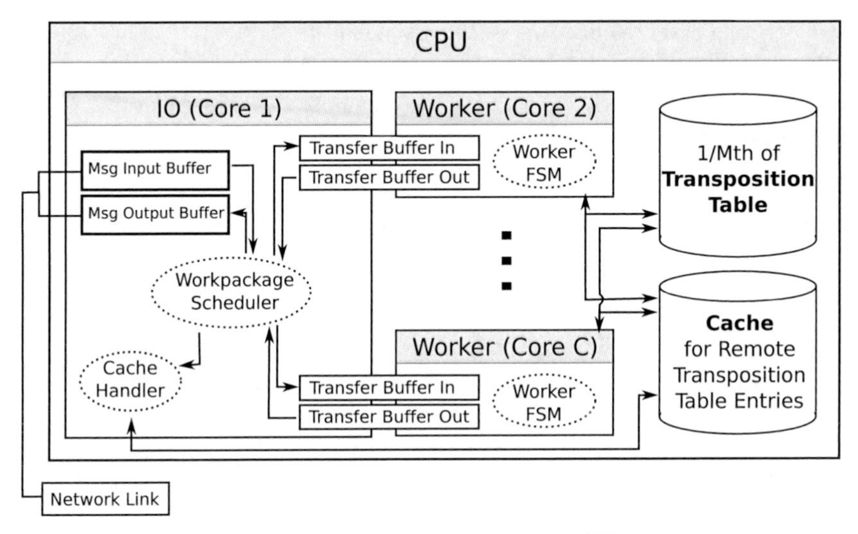

Figure 4.3: Setup of an MPI rank on a CPU.

4.2.4 Implementation Details

We concentrate on homogeneous HPC systems with a fast, low-latency Infiniband interconnect consisting of N compute nodes (CNs), each having P CPUs that in turn each have $C > 1$ compute cores that share a CN's entire memory. We assume this is a model that fits modern HPC cluster systems. We use MPI for message passing and assign one MPI rank, thus one instance of our program, to each CPU. Hence, we will run $M = N \cdot P$ program instances in parallel[4]. We devote one core (IO) of each CPU to a single thread handling message passing and work package distribution, while the remaining $C - 1$ cores (workers) of each CPU are bound to worker threads. Figure 4.3 illustrates this setup of one MPI rank on a CPU.

The IO and worker cores communicate using ring-buffers (Transfer Buffer In/Out in the figure) that reside in shared memory. The main loop running on the IO core reads available messages containing work packages from the network link and stores a reference to the according memory location in a buffer. Afterwards, a work package scheduler distributes the received packages among the workers' ring-buffers, balancing the work load. Workers poll the transfer buffers and start computation once they find a package and, if required, send response messages back to the IO core using the corresponding buffers. In turn, the IO core period-

[4]In practice, assigning one MPI rank to each CPU is important on most systems to allow for faster and more homogeneous memory access times as the contention at, e.g., QPI links can be reduced significantly.

ically collects messages from the workers' ring-buffers and forwards them to the network link appropriately[5].

Although ring buffers between dedicated communication partners can be implemented lock free and thus allow for efficient shared memory communication, in our experiments it turned out that, allowing workers to look for work packages not only in their own but also in other workers queues, improves the system's overall work load, even though the ring-buffers require explicit locking in this case.

All worker cores need access to the part of the distributed transposition table (cf. Section 4.2.3) that is associated to the MPI rank, represented by the upper cylinder and corresponding arrows on the right in Figure 4.3. The duplicated statistics of frequently visited tree nodes (cf. Section 4.2.2) are stored in a second local transposition table, denoted with *Cache* in the figure. The transposition table cache is updated not only by workers but also by the IO core. The IO core receives the data of new transposition table entries from remote ranks and takes entries from the local transposition table to send them to remote ranks for their duplication. Those tasks are handled by the cache handler that resides on the IO core.

During search, a number of simulations are computed in parallel. We denote this number with S_{par}. Each of S_{par} simulations running in parallel suffers loss of information represented by the results of the $S_{\text{par}} - 1$ other simulations that would be available in a sequential UCT version. Obviously this impairs the search quality [89] and urges us to keep S_{par} as small as possible.

In total the algorithm requires us to determine:

- An overload factor O to compute the number of simulations that run in parallel on M MPI ranks using $C - 1$ worker cores each: $S_{\text{par}} := (C-1)MO$. It should be sufficiently large to effectively overlap communication with addition simulation computations.
- A policy for duplication and then also synchronization of frequently visited tree nodes.

While the value of the overload factor O can be determined empirically, we discuss and propose a policy for tree node duplication and synchronization in Section 4.2.2. Note that Distributed-Tree-Parallelization may be configured to behave comparably to Root-Parallelization by sharing near-root nodes immediately. On the other

[5]For practical implementations, care must be taken to pass around references to corresponding memory locations rather then copying entire messages whenever possible. We recommend the allocation of a large memory buffer using the MPI function `MPI_Alloc_mem` at program start and handling the memory management within the user application, as frequent allocation and freeing of dynamic memory can easily lead to unpredictable, highly inhomogeneous processing times.

extreme, Distributed-Tree-Parallelization behaves like Tree-Parallelization if tree nodes are never duplicated.

The search process begins by sharing the root node among all MPI ranks. Then, each rank starts S_{par}/M simulations. The data structure representing a simulation consists in principle only of the state-action stack containing all states visited and actions taken during simulation. Together with each state, we store the MPI rank where the action selection took place. The rank information allows for determining the destination ranks for necessary UPDATE messages. For UPDATE messages we additionally include the playout's reward and in our Go specific implementation also some further information about the terminal position to support some domain specific MC-heuristics.

4.2.5 Load Balancing

We assume the playout work packages to make up by far the largest portion of the overall computational load that comes along with each single simulation. By design of the algorithm, these work packages are distributed randomly across all worker MPI ranks. This, however, does not ensure that work is well balanced at any time but only on average. Even further, the computation of a playout is not strictly bound to a fixed MPI rank as the transposition table does not need to be accessed. We make use of this absence of data dependency and forward playout computations to other ranks in case we notice *unusual high workload* on the rank that originally had to compute it. To perceive unusual high workload at a specific rank, we look at the number of MOVE messages that reside in the rank's message buffers. Being aware that in total at most S_{par} simulations are running in parallel, we expect to find an average number of (S_{par}/M) MOVE messages. We define a rank's workload to be unusual high in case we find more than $(S_{\text{par}}/M) \cdot K$ with $K > 1$ MOVE messages in the rank's message buffers. Playout work packages are then forwarded to other, randomly chosen MPI ranks that might of course themselves forward the work again until a rank with appropriate workload was found. Choosing $K > 1$ ensures that such CNs exist. In our experiments we achieved satisfying results with $K = 1.2$.

Taking the distributed simulations, the duplication and synchronization of frequently visited tree nodes and the playout redistribution for load balancing together, our approach effectively combines all major parallelization techniques presented in Section 4.1.1. Depicting our approach in the manner we did it for the other MCTS parallelizations above, we obtain Figure 4.4.

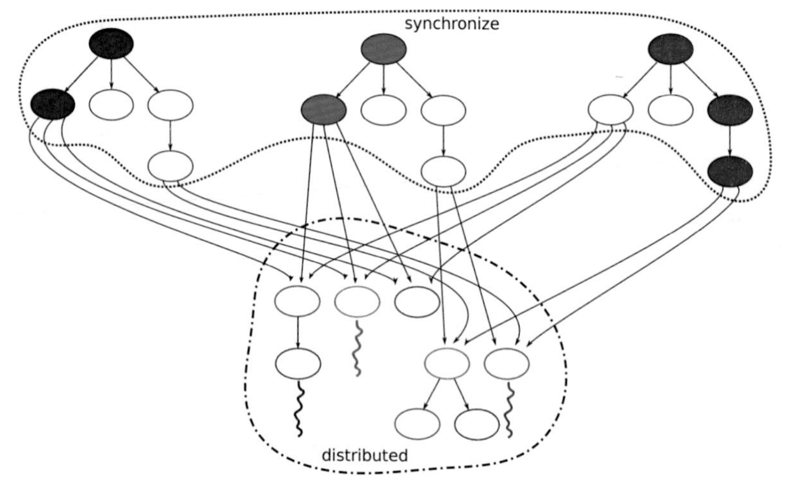

Figure 4.4: An illustration of our Distributed-Tree-Parallelization. Frequently visited nodes near the root of the tree are duplicated on each compute node. Those shared tree nodes are frequently synchronized. By far most of the tree nodes, that are less frequently visited, are stored in a distributed transposition table. Computational tasks are performed were the corresponding data is located. Consequently, simulations become distributed as well and may need to jump between different compute nodes during computation. Different colors represent computations performed by different compute nodes.

4.2.6 Broadcast Nodes

A bottleneck in the parallelization approach as described so far, is the need for regular all-to-all communications to synchronize the statistics of duplicated near-root nodes. As current MPI libraries still lack efficient, asynchronous many-cast operations, we add additional CNs to help spreading synchronization messages[6]. In the remainder of this thesis, we denote the MPI ranks that handle the tree search with *search ranks* and call those helping with shared tree node synchronization *broadcast ranks*. We empirically determined a good ratio to be one broadcast rank for every four search ranks. Hence, each search rank has a dedicated broadcast rank to send all synchronization messages to. In turn a search rank will also receive synchronization messages of the originating search ranks only via this dedicated broadcast partner. The broadcast ranks themselves synchronize by all-to-all communication.

The design choice of using broadcast ranks even allows for merging synchronization messages originating from different search ranks that affect the same tree node and hence, greatly helps to reduce network traffic and takes over additional work load from the search ranks. To further support this effect, messages are artificially delayed by the broadcast ranks to increase the probability that messages containing information for equal tree nodes can be merged. This delay and merge technique considerably reduces the number of synchronization messages that have to be received and processed by each single search rank. Otherwise, the number of those messages might double with each doubling of the search ranks and thereby limit the scalability of the overall algorithm.

We determined the broadcast to search ranks ratio (1:4) empirically for 16 search ranks. For other numbers of search ranks, different ratios might yield better performance. Furthermore, our broadcast rank's implementation is single threaded. Hence, further performance improvements might be obtainable by handling work packages like synchronization message duplication and merging on additional compute cores. In summary, our implementation of broadcast ranks is rather a proof of concept than being optimal.

The use of an Infiniband interconnect allows for so called RDMA (Remote-Direct-Memory-Access) communication, that helps shifting communication related work almost entirely to the corresponding Infiniband hardware. This technique not only provides very low latencies (less than 1 microsecond) but simplifies overlapping of communication with computation remarkably. A practical drawback of RDMA, however, is the need of preallocated receive buffers at each MPI rank for each of

[6]The use of blocking many-cast operations like `MPI_Allreduce`, that might be used for Root-Parallelization, is inviable for our approach, because it would block the IO core. Blocking an IO core would immediately lead to blocking distributed simulations processing and thereby might harm the whole system's performance considerably.

the rank's potential communication partners (peers). Even further, these buffers need to be polled by the CPU in order to recognize the completion of asynchronous message receive operations. Hence, in terms of memory requirements and polling time (and thus CPU time), the use of RDMA in combination with all-to-all communication does not scale to arbitrarily sized clusters. Actually, OpenMPI limits the number of RDMA peers per default to only 16 per MPI rank but allows for adjusting this number.

To address this issue, we introduce a communication pattern for the broadcast ranks that makes each broadcast rank communicate with a limited number of L other broadcast ranks in addition to its associated search ranks. The pattern minimizes the number of hops and targets at uniformly distributing the communication and work load among the broadcast ranks. Recalling that the number of search ranks is denoted by M, and having $B = M/4$ broadcast ranks, we determine positive numbers k, n and $m \in \mathbb{N}$ that fulfill the following two conditions:

$$k^n \cdot m = B \tag{4.8}$$

$$n + m - 1 \leq L. \tag{4.9}$$

In case more than one realization of k, n and m can be found, we are interested in the one minimizing $k + n$. We can then define our logical broadcast network as m connected k-ary n-cube networks, where each node belongs to one cube and has links to its counterparts in the $m - 1$ remaining cubes in addition to n links to the neighboring cube nodes (one in each dimension). We label the broadcast ranks by tuples (a, b) with $a \in \{0, \ldots, m-1\}$ and $b \in \{0, \ldots, k-1\}^n$. Here, a determines the k-ary n-cube the rank is part of while $(b_0, \ldots, b_{n-1})$ represents the index of the rank in the respective cube. Inside the cube we allow rank (a, b) the forwarding of messages to its successors $(b_0, \ldots, (b_i + 1) \mod k, \ldots, b_{n-1})$ in each dimension $i \in \{0, \ldots, n-1\}$. In addition, each rank (a, b) can forward messages to its respective counterparts in the other cubes, i.e., to all ranks (x, b) with $x \neq a$, resulting in a total of $n+m-1$ outgoing links for each broadcast rank. The number of incoming links is equal, although the cube related ingoing and outgoing links differ for $k \geq 3$. Procedure Broadcast shows in pseudo-code how a broadcast is performed in the network.

Figure 4.5 depicts an example of the logical interconnect for 16 compute nodes and a peer limit of $L = 6$. Figures 4.5b to 4.5d show the 3-hop broadcast operation that can logically be split into a 2-hop binary 2-cube (i.e. hypercube) broadcast operation and 2^2 associated clique communications between the single 2-cube nodes and their respective counterparts. The maximum number of hops needed for each message is equal to $n(k-1)+1$ while each broadcast rank receives messages from at most L other broadcast ranks as desired.

Procedure Broadcast((a^*,b^*), dimension)

Data: $k, n, m \in \mathbb{N}$ define m connected k-ary n-cubes. Nodes are indexed by tuples $(a, b_0 b_1 \dots b_{n-1}) \in \{0, \dots, m-1\} \times \{0, \dots, k-1\}^n$. dimension $\in \{0, \dots, n-1\}$. (a, b) is the current node index and (a^*, b^*) denotes the index of the node that initiated the broadcast.

input : Index (a^*, b^*) of the node that initiated the broadcast and dimension.

```
1  if (a,b) = (a*,b*) then
2  |  dimension ← n;
3  end
4  for 0 ≤ x < m and x ≠ a do
5  |  forward message to node (x,b);
6  end
7  for 0 ≤ i < dimension do
8  |  if ((b_i + 1) mod k) ≠ b*_i then
9  |  |  new_dimension ← i;
10 |  |  if ((b_i + 2) mod k) ≠ b*_i then
11 |  |  |  new_dimension ← i + 1;
12 |  |  end
13 |  |  forward message to (a, b_0 ... ((b_i + 1) mod k) ... b_{n-1}) and
14 |  |  call Broadcast ((a*,b*),new_dimension) there on message arrival;
15 |  end
16 end
```

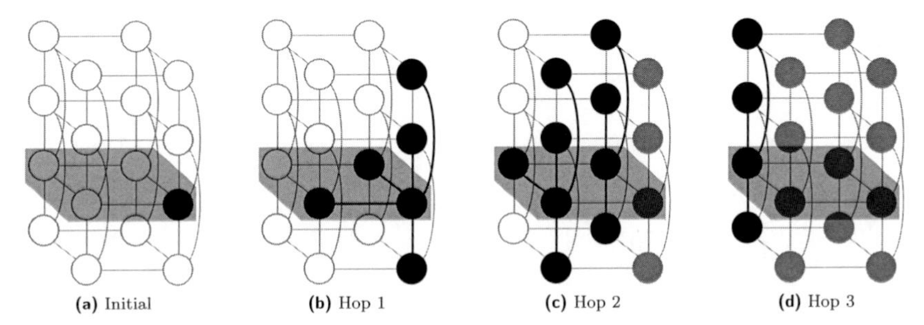

(a) Initial (b) Hop 1 (c) Hop 2 (d) Hop 3

Figure 4.5: Example 3-hop broadcast operation with 16 broadcast ranks and a maximum of 6 peers per rank, giving $(k, n, m) = (2, 2, 4)$. The gray box marks the 4 ranks that are used in the binary 2-cube (i.e. hypercube) communication.

4.3 Experimental Results

In this chapter, we present the experimental setup and results achieved with our Go engine Gomorra that incorporates the Distributed-Tree-Parallelization introduced above. We start with experimental results concerning the scalability of our parallelization in terms of search quality that were produced by having Gomorra play against itself and the open-source Go engines FUEGO[7] version 1.1 and Pachi[8] version 10.00 (Satsugen). Then we present more detailed insights by showing the development of achieved simulation rates, measured workloads and network bandwidth usage for varying numbers of MPI ranks. Whenever provided, confidence intervals are given with a 95% confidence level. In addition, we analyze the impact of increasing numbers of parallel simulations and MPI ranks on the playing strength separately, to provide a more complete picture of the algorithms behavior. Furthermore, we measured the runtime of certain OpenMPI library calls for varying numbers of CNs in order to estimate the systems behavior in terms of scalability when using even more compute resources. We end up with an empirical analysis of the frequencies and overheads of the distinct work packages in the sequential and distributed program version and measured the bandwidth requirements imposed by different parts of the parallel algorithm in order to determine possible future bottlenecks.

4.3.1 Setup

Our computer Go engine Gomorra implements several state of the art enhancements over basic UCT and proved its playing strength previously in several games against the currently strongest computer Go programs. In our experiments different instances of Gomorra play against each other on a 19×19 board size, giving each player 10 minutes to make all moves in a game if not stated otherwise. The distribution of the time used per move computation varies among different phases of the game. Gomorra's time management is loosely based on [52]. We choose the following values for the Distributed-Tree-Parallelization parameters: $N_{\text{dup}} := 8192$, $N_{\text{sync}}^{\min} := 8$, $N_{\text{sync}}^{\max} := 16$, $\alpha := 0.02$ and $O := 5$. Note that the optimal values will depend on parameters of the compute resources such as network latency and bandwidth as well as on the ratio of processor speed to work package size. Although reducing N_{dup} decreases the communication overhead for single simulations because less CN hops take place, the overhead for synchronizing shared nodes statistics increases because more nodes are shared. However, few hops per simulation allow for keeping O small. Furthermore, lower values of $N_{\text{sync}}^{\min}$

[7] The open source Go playing program FUEGO is available online: `http://fuego.sourceforge.net/`
[8] The open source Go playing program Pachi is available online: `http://pachi.or.cz/`

Table 4.1: The actual numbers m, n and k used for the m-connected k-ary n-cube logical broadcast network in the experiments.

bc-ranks	m	k	n
1	1	1	1
2	2	1	1
4	4	1	1
8	8	1	1
16	8	2	1
32	8	2	2
64	8	2	3

and $N_{\mathrm{sync}}^{\mathrm{max}}$ lead to increased network traffic as more synchronization messages are sent.

For the broadcast operations performed between the broadcast nodes, Table 4.1 shows the configuration parameters of the logical communication network for all numbers of broadcast ranks (bc-ranks) used during our experiments. We chose this values with the objective to keep the number of communicating peers for each broadcast rank below 16 while minimizing the number of required hops for each broadcast operation[9]. As each broadcast rank communicates with up to 4 search ranks as explained in Section 4.2.6, 12 peers remain for inter-broadcast nodes communication. Hence, it must always hold that $n + m - 1 \leq 12$.

For our experiments we use up to 160 CNs of the OCuLUS cluster[10], each one equipped with 2 Intel Xeon E5-2670 CPUs (16 cores in total) running at 2.6 GHz and 64 GByte of main memory. The CNs are connected by a 4xQDR Infiniband network. We use OpenMPI (version 1.6.4) for message passing.

4.3.2 Performance and Scalability

In this section we examine our parallelization's overall performance characteristics by having Gomorra play games against itself as well as against the open-source Go programs FUEGO and Pachi with varying numbers of compute nodes. Like Gomorra, FUEGO and Pachi are MCTS based Go programs. Pachi was configured to play on a single CN using 16 cores with identical time restrictions as Gomorra. With this configuration, Pachi computed about 30,000 simulations per second on the empty 19×19 board. As FUEGO is substantially weaker than PACHI, we relieved it from the time restrictions and configured it to compute a large constant

[9]OpenMPI version 1.6 recommends 16 as the maximum number of peers to use for eager RDMA communication.
[10]The system specification of the OCuLUS cluster is available online: `http://pc2.de/`

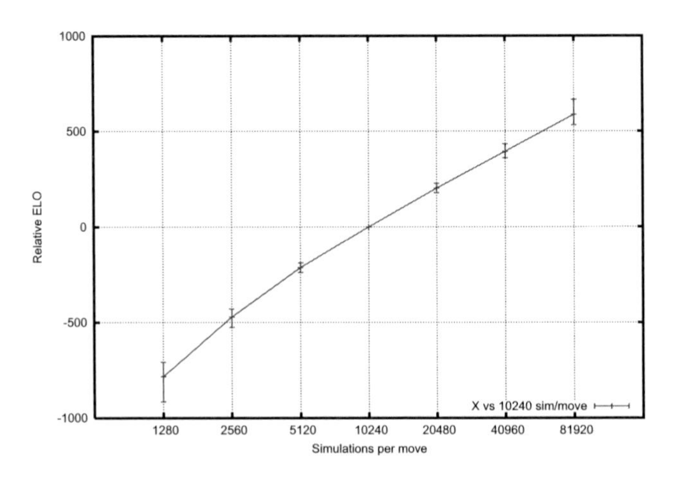

Figure 4.6: Evaluation of Gomorra's playing strength when playing against itself on a single compute node.

amount of 250,000 simulations per move decision. We hereby leverage FUEGO's playing strength to be comparable to the strength of Gomorra and Pachi.

The most important measure for the performance of a parallel MCTS algorithm and its scalability is the development of playing strength with an increasing number of CNs. We measured the playing strength in ELO[34]. A win rate of $p \in (0, 1)$ of one program instance over the other, translates to a relative ELO value of $\log_{10}(p/(1-p)) \cdot 400$. Figure 4.6 shows this development of playing strength for the sequential version of Gomorra when playing with varying number of simulations per move. Figure 4.7 shows the according graph for the parallel version when using varying numbers of CNs. All games played for these figures were games against Gomorra itself. The reference configuration for the sequential version computed a fixed number of 10,240 simulations per move. The reference configuration for the parallel version used 8(+2) CNs. Here, we write 8(+2) for 8 search CNs (i.e. 16 search ranks) and 2 broadcast CNs (i.e. 4 broadcast ranks). As we can see, the sequential version can improve by more than 1000 ELO from 6 doublings of the number of simulations performed per move whereas the parallel version improves by about 700 ELO from 6 doublings of the number of CNs. In order to relate both curves and to understand why the parallel version does not also improve by more than 1000 ELO, we first want to know if 6 doublings of CNs

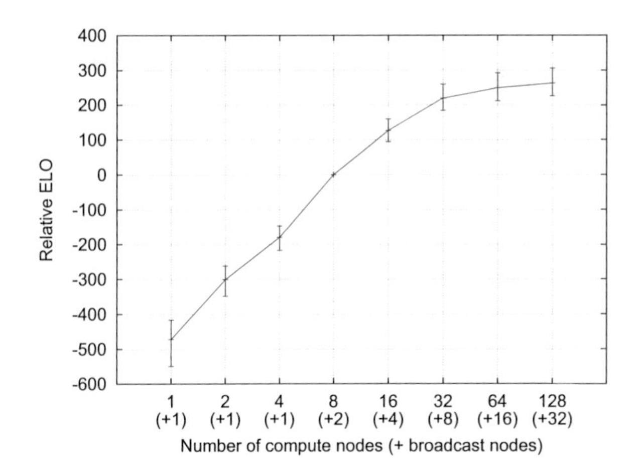

Figure 4.7: Evaluation of Gomorra's playing strength when playing against itself using Distributed-Tree-Parallelization for varying numbers of compute nodes.

also results in 6 doublings of the number of simulations that are computed for each move decision. Therefore, we look at the simulation rate, i.e., the number of simulations that can be computed per time unit for varying numbers of CNs.

Figure 4.8 shows the scalability of the simulation rate and Figure 4.9 the development of the average workload measured at the worker cores. Both diagrams show 4 curves for measurements at different phases of the game: At moves number 10, 50, 100 and 150. We plot curves for different game phases, because the playouts become shorter in later game phases and this impacts the distribution of computational load among the different work packages of each single simulation. We can see that the number of simulations scales uniformly for up to 32(+8) CNs and observe a slightly degraded scalability for more nodes. Figure 4.9 in contrast shows a more remarkable drop of measured workloads when using more than 32(+8) CNs. This observation also matches the reduction of playing strength scalability observed in Figure 4.7.

Observing decreasing workloads leads to the assumption that either the distribution of work packages or the synchronization of shared tree nodes or both becomes more challenging with increasing number of CNs. In order to explain the observed numbers, we examine the actual communication related requirements in terms of used bandwidth and computational overhead.

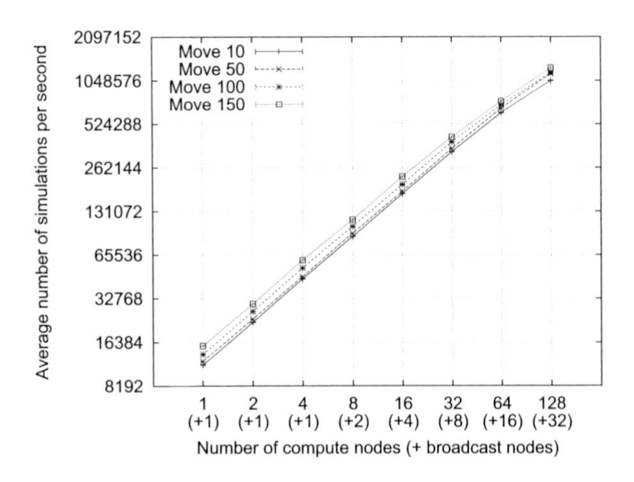

Figure 4.8: Scalability of simulation rate with increasing number of CNs at different game phases.

Figure 4.10 shows the average bandwidth usage measured on a worker MPI rank for outgoing messages. The corresponding measures for broadcast ranks are shown in Figure 4.11. In both cases, the required bandwidth remains well below the available bandwidth for any configuration. Using 4xQDR Infiniband as in our case, the theoretically available bandwidth per CN is about 32 Gbit/s. Recall that each CN is used by two MPI ranks, hence the total amount of bandwidth available per MPI rank is about 16 Gbit/s. Accordingly, the bandwidth requirements stay well below their limits.

In order to estimate the computational overhead related to communication on worker and broadcast CNs, we measure the average message loop iteration time, as most of the communication related computation time is located in this loop. In fact, we assigned a whole core solely for running this loop. Also computations performed by the MPI library are bound to this core and hence, will inherently be included in our measurements. Some of the communication related work, however, is not included as some parts, such as the composition of messages, were shifted to the worker cores.

Figure 4.12 shows the aforementioned measurements of the main message loop's average turnaround time on the worker CNs with varying number of CNs. During each iteration, the main message loop calls the MPI function `MPI_Testsome` once

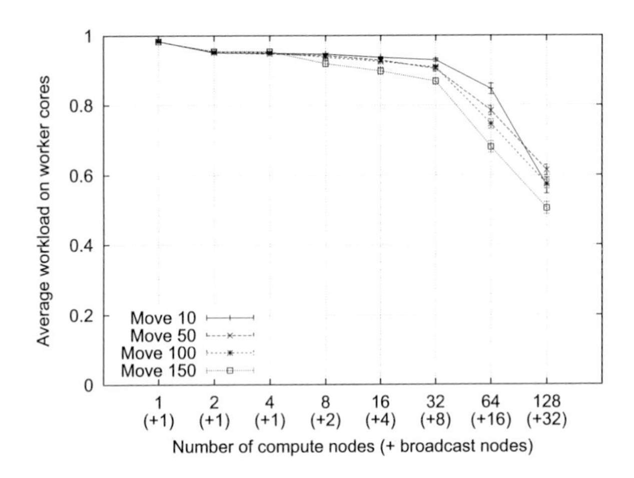

Figure 4.9: Average workload at worker cores with increasing number of CNs at different game phases.

in order to get informed about newly received messages and about the completion of pending asynchronous send operations. In addition it collects all messages supplied through the worker cores' transfer buffers (cf. Figure 4.3) and asynchronously sends them to the corresponding destination ranks. Furthermore all received simulation related messages are forwarded to the corresponding workers' transfer buffers. Additionally, incoming duplication and synchronization messages for shared tree nodes are immediately processed and corresponding synchronization messages are asynchronously sent to the MPI rank's associated broadcast rank, represented by the cache handler in Figure 4.3. Figure 4.12 contains an additional curve for the average time spend inside the `MPI_Testsome` routine, that consumes by far most of the time. This stems from the fact that OpenMPI uses the call to `MPI_Testsome` not only to poll message buffers to recognize completed RDMA communications but also for other library internal work. We can see that `MPI_Testsome` is responsible for steadily increasing turnaround times of the main communication loop with increasing numbers of CNs. This is likely the case because RDMA communication requires the MPI library on each rank to provide a number of receive buffers for each peer that regularly sends messages to this rank, and to poll some memory locations associated to those buffers in order to recognize the completion of incoming messages. OpenMPI allows for restricting the use of RDMA communication to a limited number of peers to bound the polling time and

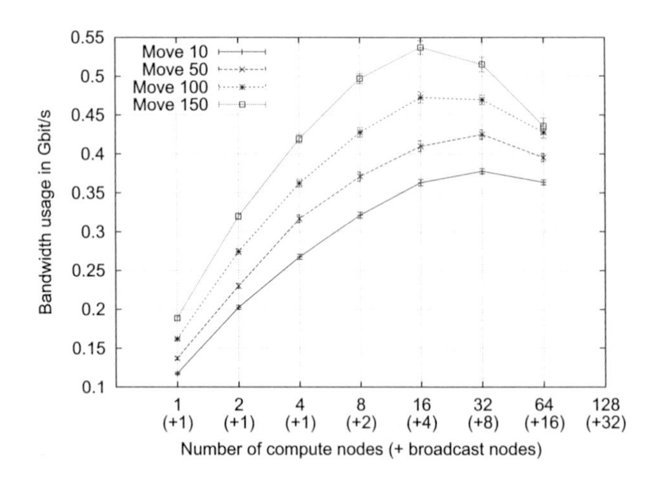

Figure 4.10: Average outgoing bandwidth usage at search ranks with varying numbers of computer nodes.

the memory consumption for the receive buffers. However, limiting this number comes with a significant increase of the communication latency and the protocol for non-RDMA communications might require a CPU involvement a number of times in the course of each single message transfer. This in turn impairs the overlapping of communication with computation, as communication can only progress when we explicitly assign execution time to the MPI library (e.g. by a call to `MPI_Testsome`). Our experiments showed that the aforementioned drawbacks of partly non-RDMA communication heavily outweigh the increasing polling time and memory consumptions when using RDMA for communicating with all peers. Obviously, for even larger systems the use of RDMA communication has to be restricted.

It is important to understand how valuable a number of parallel simulations are in comparison to an equal number of sequentially computed simulations. As our approach allows for running an arbitrary number of simulations in parallel on a given number of cores, we measured the relative differences in playing strength when playing with Gomorra using varying numbers of parallel simulations against a version of Gomorra that always uses the same degree of parallelism, i.e the same number of parallel simulations. Precisely, we configured Gomorra to make exactly 10 simulations in parallel at a time as the reference version and made

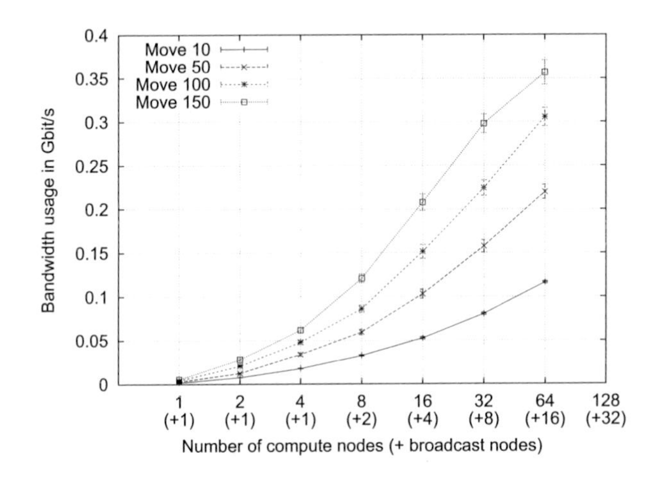

Figure 4.11: Average outgoing bandwidth usage at broadcast ranks with varying numbers of computer nodes.

it play against versions of Gomorra that made exactly 20, 40, 80, 160, 320, 640 and 1280 simulations in parallel. For all pairings, we played 1000 games where each instance of Gomorra had to make exactly 1280, 2560, 5120, 10240, 20480, 40960 and 81920 simulations per move. The results are shown in Figure 4.15. We observe the expected decrease in playing strength for increasing degrees of parallelism and equal number of simulations. Somewhat less intuitive is the fact that high degrees of parallelism do not impair the search quality a lot for lower numbers of simulations per move. In fact, this is likely an effect of Progressive Widening (cf. Section 3.3.3), that initially restricts the search to only a very few moves following a heuristic move prediction. Hence, for low numbers of simulations per move, we have limited ability for making heavily wrong decisions. We can also see that, at the long end, the search quality impairing becomes less significant the more simulations are performed per move in relation to the number of simulations that are computed in parallel. Similar experiments were already conducted by Richard Segal in his paper about experiments with the Go program FUEGO regarding the scalability of parallel MCTS [89] by simulating Tree-Parallelization for very large shared memory machines. Segals experiments show a comparable development of the measured relative strength's with longer thinking times, i.e., for higher numbers of simulations performed per move. Hence, it appears to be the

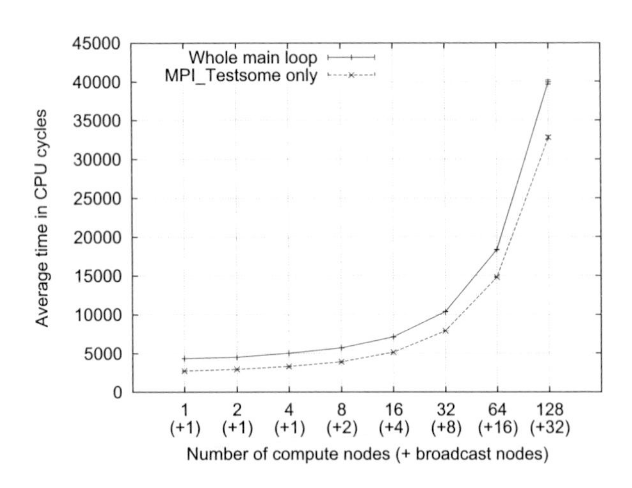

Figure 4.12: Time requirement of MPI_Testsome in comparison with the entire message main loop on search ranks.

case that the impairing due to higher degrees of parallelism continuously decreases with longer thinking times and likely becomes almost negligible.

Finally, in order to make sure that the scalability of our parallelization in terms of playing strength is not restricted to self play experiments, we measured the playing strength development with varying numbers of CNs against the open-source Go engines FUEGO and Pachi. Figure 4.13 shows the results for FUEGO. Also against FUEGO we measured a significant improvement in playing strength when using more CNs for up to 32(+8) CNs. However, gaining about 400 Elo, the improvement is considerably less than in self-play where we achieved more than 700 Elo (see Figure 4.7). Improvements over modified versions of the same Go program are known to be generally better than improvements over other Go programs. Effects comparable to our measurements were also observed and published by, e.g., [9]. Even further, we see that the drop of workload and the degrading simulation rate scalability leads to a more severe impairing for 64(+16) CNs and more. We assume this impairing that actually leads to a drop of playing strength for large systems, originates in higher ramp-up times for non-self-play games. In self-play games, it is likely that both program instances are searching most of the time in the same portion of the search space as both instances use, e.g., the same heuristics and playout policies. This increases the chance of being able to

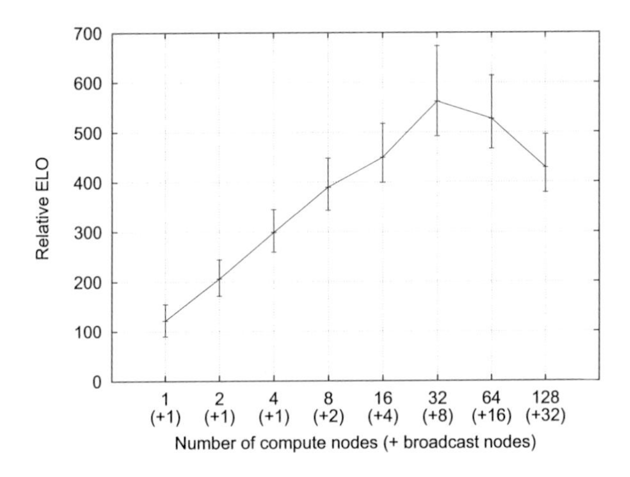

Figure 4.13: Evaluation of Gomorra's playing strength when playing against the open-source Go program FUEGO.

reuse parts of the transposition table content of previous search runs and thereby leads to reduced ramp-up times. This is especially true in case the root node was already duplicated as all search ranks can then immediately start simulations. Figure 4.14 shows the corresponding results for Pachi. Also here, Gomorra gains playing strength for up to 32(+8) CNs and shows a slight drop when using more CNs.

To get an idea about the advantage our parallelization has from spreading a single search tree representation over doing rather independent searches on each rank but sharing near root nodes (i.e. Root-Parallelization), we configured Gomorra to use exactly the same way of sharing and synchronizing of near root nodes but building up otherwise independent search tree representations on each rank. The results are shown in Figure 4.16 for self-play experiments and in Figure 4.17 for experiments with FUEGO[11]. Again, the improvement by the use of more CNs is lower when playing against FUEGO than in self-play. But even more we can see a rather big drop of playing strength when compared to our proposed parallelization Distributed-Tree-Parallelization. Mind that we run two MPI ranks on each CN, explaining the different measurements already for a single CN. This should not be

[11]Note that these experiments were conducted with an older version of Distributed-Tree-Parallelization that were given only 5 min per game (instead of 10 min), explaining a generally lower playing strength compared to the results presented in Figure 4.13

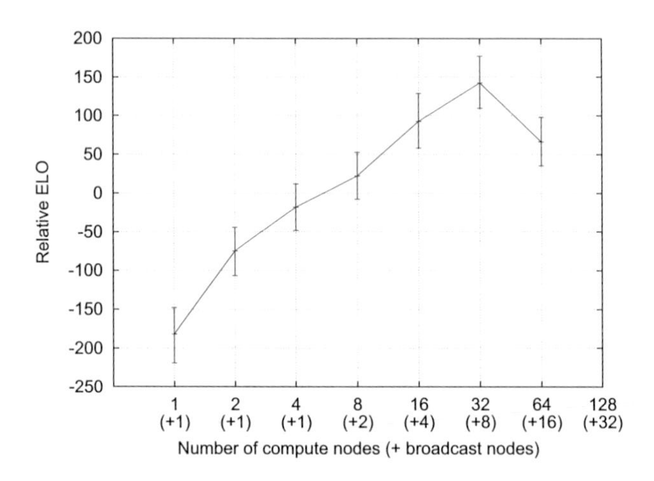

Figure 4.14: Evaluation of Gomorra's playing strength when playing against the open-source Go program Pachi.

understood to be a fair comparison of Distributed-Tree-Parallelization and Root-Parallelization, but should give insights about the importance of sharing the tree in our approach. Actually others, in particular [9] were able to achieve slightly better scaling with Root-Parallelization than we did. Here it is worth to mention that they explicitly tried to have the distinct CNs search in different regions of the tree, e.g., by assigning different priors to the tree nodes on different CNs, a technique called *virtual wins* in [9] (cf. Section 4.1.3). This makes sense for Root-Parallelization as it prevents having all nodes recreate almost the same search tree representation. Recall that in Root-Parallelization each CN establishes its own, rather independent search tree representation. For Distributed-Tree-Parallelization, the effect of *virtual wins* however contradicts our objective to keep the shared tree nodes as equal as possible because they represent a part of the single one shared search tree representation. Hence, observing slightly worse scaling with our Root-Parallelization version than others achieved does not contradict our expectations.

Basically we might say that Root-Parallelization as implemented and advocated by, e.g., [9] is more targeting at jumping out of or preventing to get stuck in local optima while our approach is targeting at increasing the search depth. Assuming this holds true, both parallelization methods are complementary to a certain de-

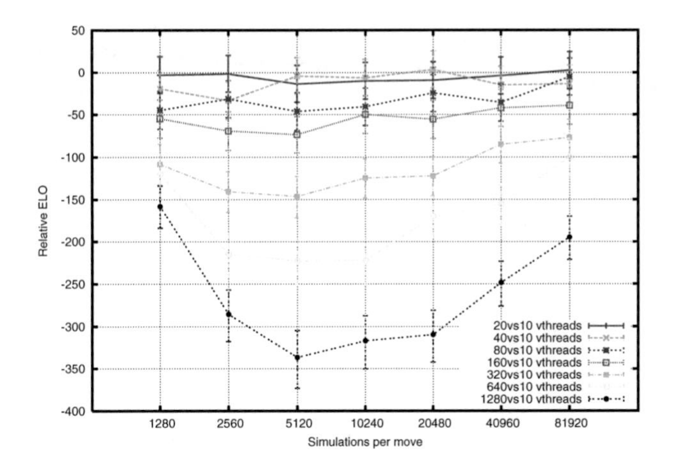

Figure 4.15: Evaluation of Distributed-Tree-Parallelization for different numbers of vthreads and varying number of simulations.

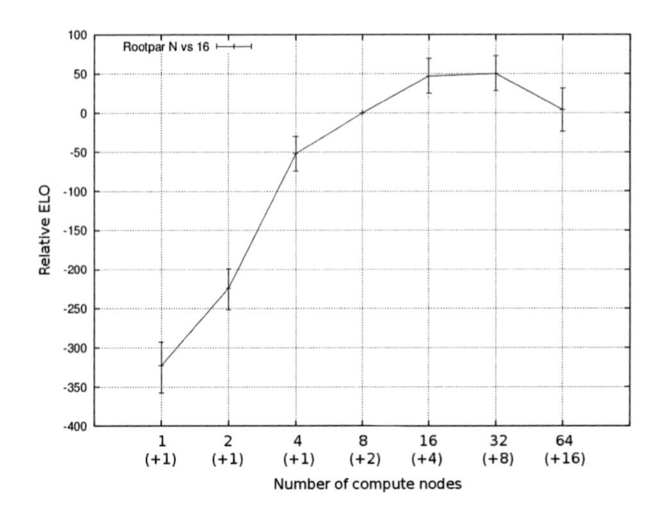

Figure 4.16: Selfplay strength scalability of Root-Parallelization

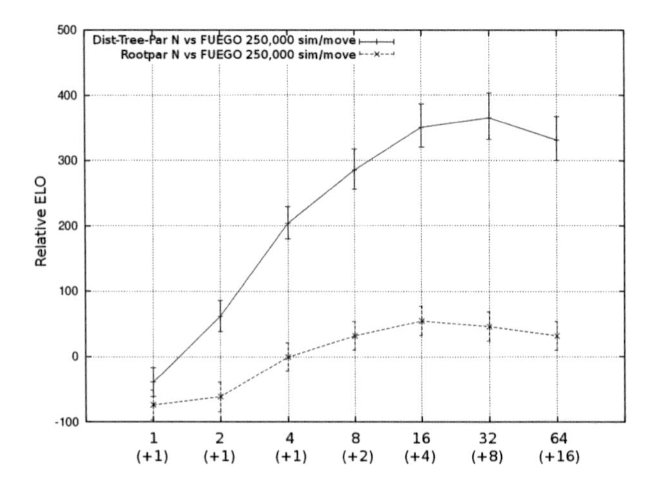

Figure 4.17: Comparing Root-Parallelization and Distributed-Tree-Parallelization playing against Fuego

gree and it might be valuable to combine both approaches by, e.g., build clusters of CNs that use Distributed-Tree-Parallelization internally while using some form of root parallelization between cluster boundaries.

4.3.3 Overhead Distributions

In this section, we present and discuss experimental results concerning the overheads of the different work packages in terms of computation time and frequency of occurrence. As a reference point, Figure 4.18 shows the average absolute computation time that is spent in the different work packages *select*, *playout*, *expand* and *backup* as they were defined in section 4.2.1, and different phases of a game, when Gomorra runs on a single core performing 10k simulations per move. In the diagram, the time requirements of the different work packages are stacked. Hence, the total amount of time spend on average for the first move is about 9.5 seconds and about 5.2 seconds for move 300. This drop in time makes completely sense, as the length of playouts, and thereby the required time for their computation, decreases to the end of the game. As the diagram clearly shows, the computation of the playouts makes up by far the largest portion of the overall computation time. An irregularity can be seen in the curve of the work package *select*. Here we

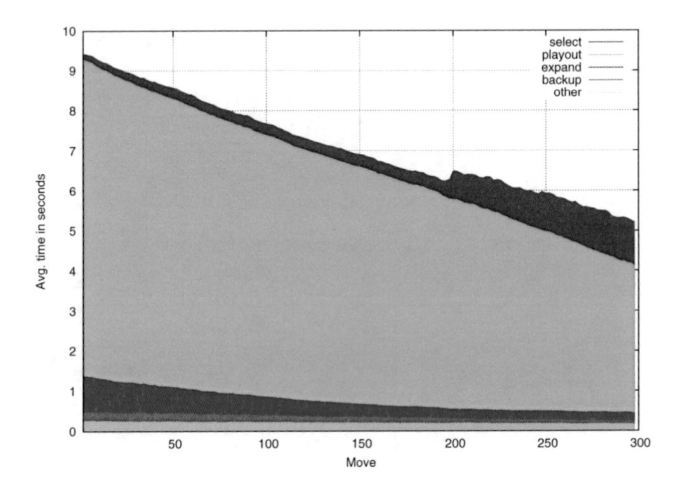

Figure 4.18: Distribution of computation time overheads in sequential version

see a huge increase of time requirement around move 200. This is due to a special behavior of Gomorra not to use Progressive Widening (PW, see section 3.3.3) after move 200. Not using PW results in directly considering all available moves at each tree node and thus makes the move selection step naturally more time consuming. The main reason for not using PW in late game phases is the fact that our move prediction system, that is the base for Progressive Widening, is trained on games that were foremost played by strong human player using Japanese rules. In territory based scoring, placing a stone on an intersection between the borders of the players' areas can be neutral (i.e. does not give points to either player). Such moves are therefore rarely played if ever. Those intersections are also called *Dame*. As a consequence, a move predictor that has been trained on games that used territory scoring, does not prioritize such moves. As most professional games are played with territory scoring, we trained our move predictor mostly with games that use territory scoring. However, filling Dames is extremely important in close endgames when using area scoring, which is typically used in Computer Go. With area scoring, filling a dame equals one point.

Figure 4.19 shows the average absolute time requirements for essentially the same work packages when running Gomorra in parallel on 16(+4) compute nodes, hence 16 worker nodes and 4 associated broadcast nodes. As each CN is equipped with 2 CPUs we actually run 32(+8) MPI ranks. This time, we configured Gomorra

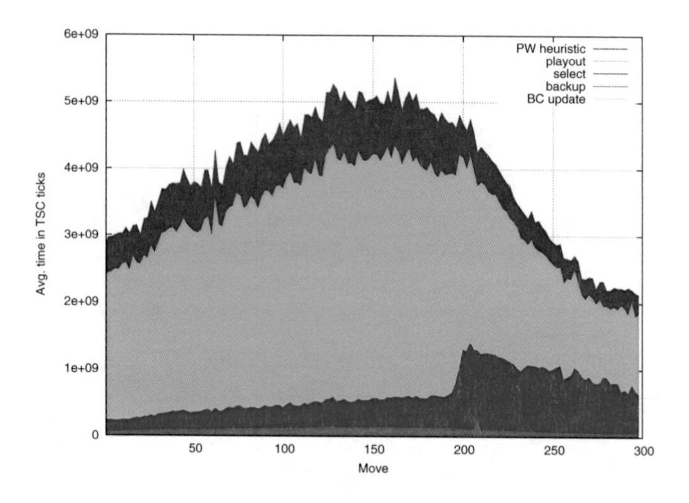

Figure 4.19: Distribution of computation time overheads in distributed version

to use a maximum of 5 minutes per game to play all its moves applying a time management as proposed in [52]. The stacked curves now have a bell shape. This is a side effect of the time management that assigns more time to the mid-game phase. Also, because of the time management, the increased amount of time spend in the selection package from move 200 onwards does not lead to an increased overall time requirement but in fact reduces the number of simulations performed. In addition to the work packages we examined with the sequential version of Gomorra (cf. Figure 4.18), we also explicitly plotted the time requirement for the heuristics' computation that forms the input for Progressive Widening and Progressive Bias. The time for the heuristics' computation was included in the selection step before. We can observe an almost identical relative distribution of computation time among the work packages as already seen with the sequential version of Gomorra. In the parallel version, we must also look at the overhead generated by the need of incorporating incoming updates of shared node statistics, denoted as *BC update* in the diagram. As we can see, these additional costs are negligible when using 16(+4) CNs.

Apart of looking at the absolute time requirements of the distinct work packages, Figure 4.20 gives insights about the occurrence frequencies of the work packages. Here we can see that the computation of heuristics required by PW and PB, that takes an considerable portion of the overall processing time, actually occurs rarely.

About the same is the case for playouts. In fact, most often we handle work related to the select and backup step. We can relate the data underlying Figure 4.19 and Figure 4.20 to plot the average amount of time spent on the computation of a single package of the distinct kinds of work. The result is plotted in Figure 4.21 and, not surprisingly, shows a very unbalanced distribution of processing times for single work packages. Note that in Figure 4.21, the time axis is in log-scale in contrast to the figures before.

As already mentioned in section 4.2.5, this imbalance can significantly hurt the balance distribution of computational load. A typical simulation will have the following sequence of work packages:

1. A number of select steps that will be computed at different CNs
2. Occasionally a heuristic computation
3. A playout computation
4. One update work package for each CN visited along the way

As according to the diagrams discussed above, by far most of the time is spend during the heuristic and playout step, it is not unlikely that an MPI rank gets entirely occupied with the treatment of a number of those work packages. During this time, the rank becomes blocked for other work and might become a sink for all simulation related messages in the system. In section 4.2.5 we presented a mechanism that prevents this scenario by redistributing the playout computation as needed.

4.3.4 Effect of Parameters

We conducted further experiments to analyze the impact of UCT-Treesplit's parameters N_{dup}, $N_{\text{sync}}^{\text{min}}$ and $N_{\text{sync}}^{\text{max}}$ on the playing strength. We took the parameter settings used for the experiments presented above ($N_{\text{dup}} := 8192$, $N_{\text{sync}}^{\text{min}} := 8$, $N_{\text{sync}}^{\text{max}} := 16$, $\alpha := 0.02$ and $O := 5$) as the reference point and empirically analyzed the effect of changing a single parameter. All experimental results presented below were obtained from games between Gomorra using 32(+8) CNs and Pachi. As in the experiments presented above, each program had 10 minutes to make all its moves in a game.

Figure 4.22 shows the development of playing strength for varying values of N_{dup}. We observe a bell-shaped curve. Low values of N_{dup} result in larger portions of the tree that get duplicated. This results in more statistics that are less accurate due to delayed synchronization. For very high values of N_{dup} it happens that frequently visited tree nodes that are not yet duplicated lead to high contention

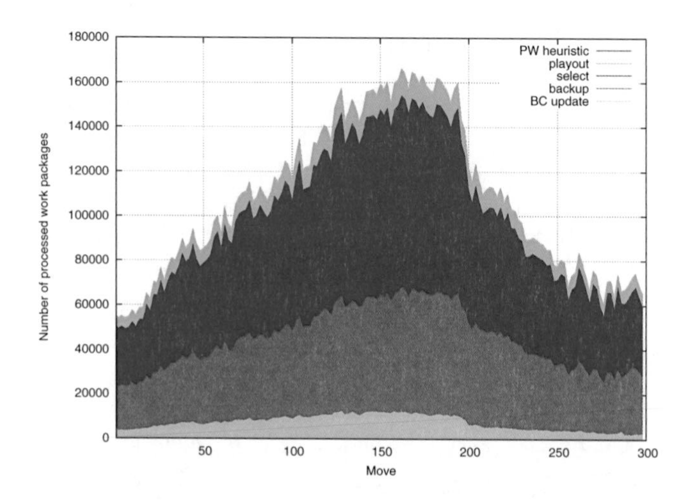

Figure 4.20: Distribution of work package occurrence in distributed version

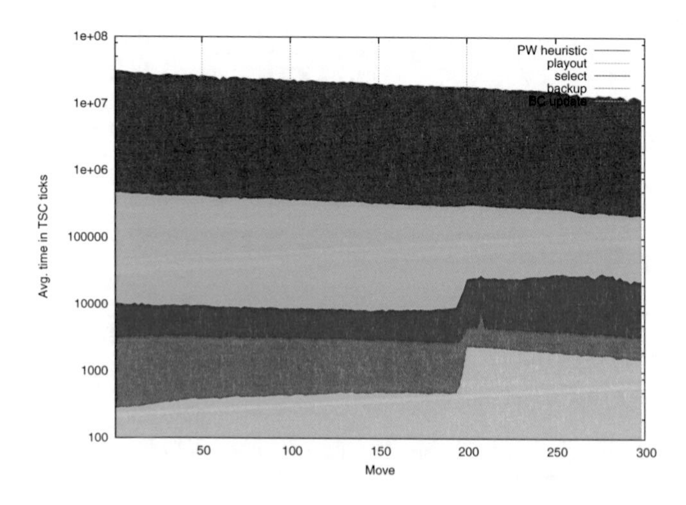

Figure 4.21: Distribution of average work package processing time in distributed version

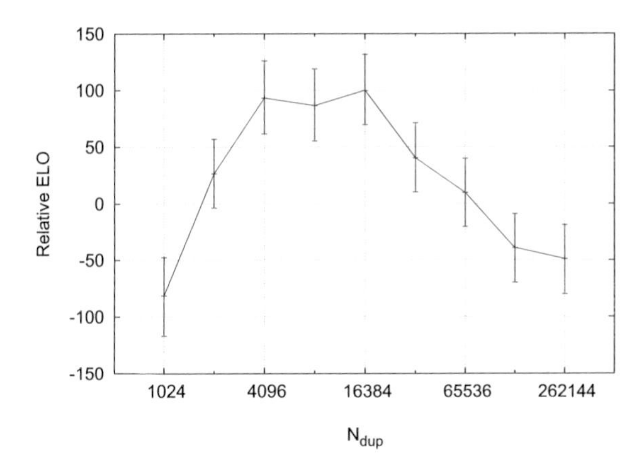

Figure 4.22: Evaluation of Gomorra's playing strength when playing against Pachi using Distributed-Tree-Parallelization and varying values for parameter N_{dup}.

on single CNs, leading in turn to severe load imbalances and thereby to reduced simulation rates.

The development of playing strength for varying values of the parameters $N_{\text{sync}}^{\text{min}}$ and $N_{\text{sync}}^{\text{max}}$ is depicted in Figure 4.23. While changing the value of $N_{\text{sync}}^{\text{min}}$ as shown on the x-axis, we set $N_{\text{sync}}^{\text{max}} = 2N_{\text{sync}}^{\text{min}}$. We see that a reduction of the synchronization rate by increasing the value of $N_{\text{sync}}^{\text{max}}$ leads to reduced playing strength. This makes sense as statistics become less accurate when increasing synchronization times. Mind that network traffic increases when synchronizing more frequently, i.e., for low values of $N_{\text{sync}}^{\text{min}}$ and $N_{\text{sync}}^{\text{max}}$.

4.3.5 Discussion of the Comparison of MCTS Parallelizations

A fair comparison of different MCTS parallelizations is challenging, if possible at all. The performance of different parallelization approaches depends on the actual hardware platform used, the specific search domain and of course the many specific implementation details that are often not presented in detail in publications.

Looking at message passing approaches, solely using a different MPI version of the same distributor might already have a severe impact on the performance of a single parallelization approach. We therefore restricted our experiments to a

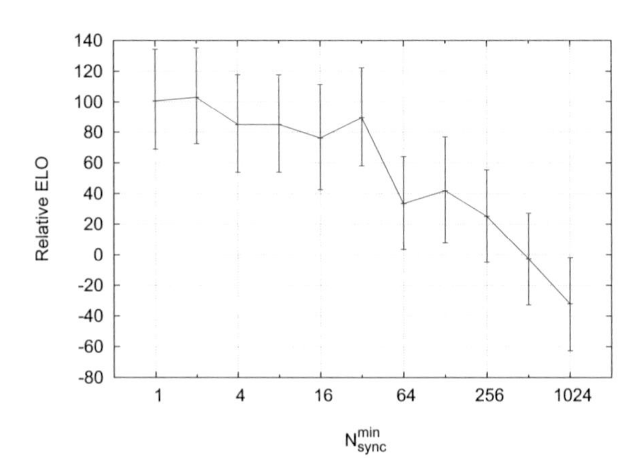

Figure 4.23: Evaluation of Gomorra's playing strength when playing against Pachi using Distributed-Tree-Parallelization and varying values for parameters $N_{\text{sync}}^{\min}$ and $N_{\text{sync}}^{\max} = 2N_{\text{sync}}^{\min}$.

comparison with one form of Root-Parallelization in order to get an idea of the gain from using a single large game tree instead of growing a number of tiny trees. While this allows us to derive a reasonable and expressive comparison of approaches, it also leaves room for more involved comparisons between Distributed-Tree-Parallelization and Root-Parallelization in general.

4.4 Chapter Conclusion

In this chapter, we presented a number of algorithmic and methodical enhancements to parallelize MCTS on distributed memory HPC systems. We developed and integrated an efficient distributed transposition table and advocate the use of additional compute nodes to support necessary all-to-all communications and thereby allow for multi-stage message merging. In conjunction with a sound policy that handles data sharing and synchronization frequencies we were able to greatly increase our parallelizations scalability. To our knowledge, Distributed-Tree-Parallelization is currently the best scaling distributed MCTS implementation. We evaluated the behavior of our parallelization in our high-end Go engine Gomorra and show that, for the game of Go, Distributed-Tree-Parallelization scales up to 128(+32) compute nodes in self-play experiments.

A close look at performance measurements regarding bandwidth usage and communication related computation overheads for varying system configurations points to possible future improvements. Presently, it appears that we reached the limits of RDMA based tiny-message communication. As one future direction, we therefore focus on limiting the number of communication peers per search rank as already done for the broadcast ranks.

In our discussion on the differences between Distributed-Tree-Parallelization and Root-Parallelization, we pointed out that, to our conviction, both approaches partly target on different objectives. While Root-Parallelization helps to prevent falling into local optima by diversifying the search on different compute ranks, Distributed-Tree-Parallelization aims on growing a single large game tree representation, leveraging deeper searches. A promising future direction might be another kind of combination of both approaches by having several instances of Distributed-Tree-Parallelization searching in parallel using Root-Parallelization.

CHAPTER 5

Move Prediction in Computer Go[1]

From the early days of research on Computer Go in the late sixties [109][110] until today, move prediction systems have been an integral part of strong Go programs. With the recent emergence of MCTS, the strength of Go programs increased dramatically to a level that appeared to be unreachable only seven years ago. MCTS learns a value function for game states by consecutive simulation of complete games of self-play using semi-randomized policies to select moves for either player. The design of these policies is key for the strength of MCTS Go programs and has been investigated by several authors in recent years [43][4][28][93][51]. Move prediction systems can serve as an important building block for the definition of MCTS policies as has been shown in, e.g., [28] and [42].

Several move prediction systems have been developed especially for the use with Computer Go [109][110][101][96][4][28] and rates for correctly predicting moves between 33.9% [4] and 34.9% [28] have been reported. However, these prediction rates have been produced under pretty unequal conditions using, e.g., different sizes of training data sets (ranging from 652 [28] to 181,000 [96] games) or different kinds and numbers of patterns (ranging from about 17,000 [28] to 12M [96]). Additionally, apart from the prediction rate, one might be interested in the technique's needs for computation and memory resources.

In this chapter, we review three prediction systems presented in literature, namely those of Stern, Herbrich and Graepel [96], Weng and Lin [103] and Coulom [28], and compare them under equal conditions for move prediction in the game of Go. Here, *equal* means that we use the same training and test data sets as well as the

[1] We presented parts of the content of this chapter before at the IEEE International Conference on Computational Intelligence and Games (cf. [104]).

same set of shape and feature patterns to represent board configurations and move decisions. To that end, we had to modify the algorithm of Stern et al. in a way already proposed by Stern himself in [95] to support teams of patterns. We also slightly modified the algorithm of Weng and Lin that was not explicitly developed for the use as move predictor in Go. Apart of this, we did not investigate or argue on the theoretical plausibility of the mentioned approaches but concentrate on their empirical comparison.

The purpose of this work is the search for a time and memory efficient online algorithm for continuous adaptation of a move prediction system to actual observations which, as we believe, is a promising approach to further increase the strength of Go programs. In a larger context, the goal is to migrate knowledge gathered during MCTS search in certain subtrees into a prediction model that itself is used all over the search space by the simulation policies. This might improve MCTS's evaluation quality in complex positions. It might also help to reduce the horizon effect that occurs as a result of the limited range of vision, denoted as horizon, of tree search algorithms. Also with MCTS, near-root nodes are evaluated better than nodes further down the tree. This allows MCTS for pushing complex situations, like semeai (cf. Section 2.1.1, page 9) in case of Go, out of its range of vision. Accordingly, the horizon effect can lead to temporarily severe misevaluations.

5.1 Background

This section introduces some necessary background that is needed in the remainder of this chapter. Starting with notations, we presentation the *Bayes Theorem* and some basic techniques used with move prediction systems.

5.1.1 Terminology

As introduced in Section 3.1 we write a game as $G := (S, A, \Gamma, \delta)$, with S being the set of all possible states (i.e. game positions), A the set of actions (i.e. moves) that lead from one state to the next, a function $\Gamma : S \to \mathcal{P}(A)$ determining the subset of available actions at a state and the transition function $\delta : S \times A \to \{S, \emptyset\}$ specifying the follow-up state for a state-action pair, where $\delta(s, a) = \emptyset$ iff $a \notin \Gamma(s)$.

In the remainder, state-action pairs will be abstracted by teams of patterns with corresponding pattern-strength values. We write the vector of n pattern strength values as $\boldsymbol{\theta} \in \mathbb{R}^n$. Patterns are assumed to be binary, i.e., either they apply for a given state-action or not. We write $\boldsymbol{\phi}(s, a) \in \{0, 1\}^n$ for the binary vector denoting which of the n existing patterns apply for a given state-action pair (s, a).

The prediction systems are trained and tested on sets of game positions, each annotated with an expert's move decision that is to be learned respectively predicted. We write $D = (s, a_1, a_2, \ldots, a_n)$ for a move decision that is made up of a game position s and the list of all moves $a_1, a_2, \ldots, a_n$ available in s among which a_1 is the expert's move decision that is to be learned. We further write $\boldsymbol{D} = D_1, D_2, \ldots, D_n$ for a series of move decisions.

5.1.2 Probability Models for Paired Comparison

There exist several probability models for paired comparison. Two of the most prominent linear models are termed Thurstone-Mosteller (TM) and Bradley-Terry (BT). Given two individuals i and j and associated values θ_i and θ_j describing their respective strength, the before mentioned probability models allow to estimate the outcome of competitions between i and j. If we denote the observed strength of the individuals in a sample competition by random variables X_i and X_j respectively, a linear model for paired comparison takes the form

$$P(X_i > X_j) = P(i \text{ beats } j) = H(\theta_i - \theta_j) \quad ,$$

with H being a monotonic, increasing function, with $H(-\inf) = 0$, $H(+\inf) = 1$ and $H(-x) = 1 - H(x)$ [47]. Hence, the probability of individual i winning over j, given their respective strength values θ_i and θ_j, is assumed to be $H(\theta_i - \theta_j)$. Two different choices of H resulting in the two probability models TM and BT are introduced in the remainder.

The Thurstone-Mosteller (TM) Model

The Thurstone-Mosteller Model (TM) assumes the strength of an individual i to be Gaussian distributed with $\mathcal{N}(\theta_i, \beta_i^2)$. Thus, in addition to the strength θ_i each individual is also assigned a variance β_i^2 that models the uncertainty about the individual's actual performance. The probability that i beats j is then modeled by

$$P(i \text{ beats } j) = \mathbf{\Phi}\left(\frac{\theta_i - \theta_j}{\sqrt{\beta_i^2 + \beta_j^2}}\right),$$

where $\mathbf{\Phi}$ denotes the cumulative distribution function of the standard normal distribution.

The Bradley-Terry (BT) Model

The Bradley-Terry Model (BT) [18] is based on a logistic distribution to estimate the outcome of games between two individuals with given strength values θ_i and θ_j respectively. It models the probability that i beats j by

$$P(i \text{ beats } j) = \frac{e^{\theta_i}}{e^{\theta_i} + e^{\theta_j}} = \frac{\gamma_i}{\gamma_i + \gamma_j}, \text{ with } \gamma_i := e^{\theta_i} \quad .$$

The BT Model is also the base for the well known Elo rating system [34] that is used, e.g., to rate human chess players (cf. Section 2.1.1, page 10).

Hunter [54] derived several generalizations of the BT model. One of them allows the prediction of the outcome of competitions between an arbitrary number of teams of individuals with the simplifying assumption that the strength of a team equals the sum of the strengths of its members (note that $e^{\theta_i+\theta_j} = \gamma_i\gamma_j$):

$$P(\text{1-2 beats 4-2 and 1-3-5}) = \frac{\gamma_1\gamma_2}{\gamma_1\gamma_2 + \gamma_4\gamma_2 + \gamma_1\gamma_3\gamma_5} \quad . \tag{5.1}$$

Given a probability model for predicting the outcome of competitions between individuals or teams of individuals, we can formulate the problem of determining the individuals' strength values θ_i (in case of TM also β_i), based on a number of observations of competitions among those individuals in $D_1, D_2, \ldots, D_n$. The Bayes Theorem poses one possible way to make such inferences.

5.1.3 Bayes Theorem

From the Bayesian viewpoint, probabilities can be used to express degrees of belief. Given a proposition A, a *prior* belief in A of $P(A)$ (expressed by a probability) and a probability model $P(A|B)$ that denotes the *likelihood* of A observing some supporting data B, the Bayes Theorem relates the *posterior probability* $P(B|A)$ of observing data B given A in future experiments as follows:

$$P(B|A) = \frac{P(A|B)P(B)}{P(A)}. \tag{5.2}$$

Hence, having a probability model $p(D|\boldsymbol{\theta})$ for a move decision D given some model parameters $\boldsymbol{\theta}$, one might infer the posterior probability of the model parameters $p(\boldsymbol{\theta}|\boldsymbol{D})$ observing some move decisions $\boldsymbol{D}$ using Bayes Theorem. This inference is a basic element of the prediction systems considered in this chapter. Hence, putting the Bayes Theorem in the context of move prediction systems considered in the remainder of this chapter, it takes the following form:

$$p(\boldsymbol{\theta}|\boldsymbol{D}) = \frac{p(\boldsymbol{D}|\boldsymbol{\theta})p(\boldsymbol{\theta})}{p(\boldsymbol{D})} = \frac{p(\boldsymbol{D}|\boldsymbol{\theta})p(\boldsymbol{\theta})}{\int p(\boldsymbol{D}|\boldsymbol{\theta})p(\boldsymbol{\theta})d\boldsymbol{\theta}}. \tag{5.3}$$

A challenging point in the inference is the possible emergence of difficult to handle integrals in the computation of $p(\boldsymbol{D})$ when considering continuous distributions: $p(\boldsymbol{D}) = \int p(\boldsymbol{D}|\boldsymbol{\theta})p(\boldsymbol{\theta})d\boldsymbol{\theta}$. However, note that a point estimate of $\boldsymbol{\theta}$ that maximizes $p(\boldsymbol{\theta}|\boldsymbol{D})$ can be determined more easily, as the denominator is not of importance in this case. Maximizing $p(\boldsymbol{\theta}|\boldsymbol{D})$ gets further simplified by using a uniform prior $p(\boldsymbol{\theta})$. The inference then boils down to maximizing the likelihood $L = \prod_{j=1}^{N} P(D_j|\boldsymbol{\theta})$.

5.1.4 State-Action Abstraction with Patterns in Computer Go

A common technique for the modeling of move decisions in games is the abstraction of the state-action space with patterns that represent certain move properties. Move prediction systems can be realized by training such models on collections of move decisions. Depending on the model's level of abstraction, a corresponding prediction system can be more or less general. The more abstract the model, the better the corresponding prediction system might generalize. However, increasing levels of abstraction naturally might lead to lower prediction quality for specific cases.

Let us first consider a basic case, in that each state-action pair, i.e., each move in a given position, is represented by a single pattern. Here, each move decision can be represented in form of a list of patterns, each representing one of all possible moves. One of those patterns is marked as the final choice. A move decision is then regarded as a competition between patterns that is won by the pattern representing the final move choice. While observing such competitions, we aim to learn strength values θ_i for each of the patterns that allow the use of probability models for paired comparison as introduced in Section 5.1.2 for move prediction. Hence, a move decision yields a sequence of strength parameters vector $D = (\theta_1, \theta_2, \ldots, \theta_k)$ where θ_1 is the strength of the winning pattern[2]. A variant that allows more concise modeling is the use of pattern teams instead of single patterns for the abstraction of a move in a given position. Recall that in Section 5.1.2 we introduced the generalized BT model that can cope with teams of patterns. We extent our θ notation to teams by writing θ_{ij} for the strength value of team's i member j.

[2]For legibility reasons, we write θ_i short in this context when we actually mean $\boldsymbol{\phi}(\boldsymbol{s}, \boldsymbol{a_i}) \cdot \boldsymbol{\theta}$, i.e., the value of the element of $\boldsymbol{\theta}$ that belongs to the pattern that is active for a given state-action pair. Hereby the indexing becomes ambiguous, but it should always become clear from the context which notation is used.

In the remainder of this chapter, we distinguish two types of patterns: shape patterns and feature patterns. Both are introduced below.

Shape Pattern

Shape patterns, that describe the local board configuration around possible moves and thus around yet empty intersections were already used in the early time of research on Go [97]. Stern et al. [96] describe diamond shaped pattern templates of different sizes around empty intersections to abstract the state-action space as shown in Figure 5.1. As can be seen in the figure, they distinguish 14 different sizes, of which the largest one (14) refers to the whole board configuration. We used the same 14 types of shape patterns for all experiments presented in this chapter.

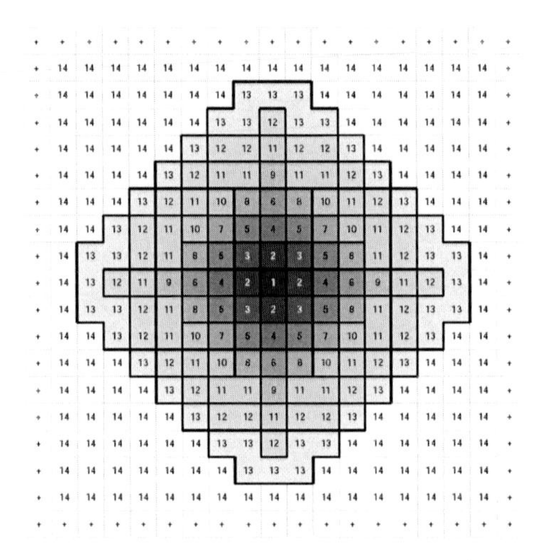

Figure 5.1: Shape-Pattern templates of sizes 1 to 14 (Source: [96]).

Note, that two shape patterns that can be converted to the same pattern by rotation and/or mirroring, can be regarded as equal in the pattern based model. Note further, that a move in the middle of the pattern will have a different value for the white and the black player in general. However, given some pattern and a corresponding value for placing a black stone in the middle obviously results in the corresponding color-inverted shape pattern to have the same value for placing a white stone in its center instead. All three operations, rotation, mirror and color-inversion can be performed with Zobrist-Hashing by simple computations on a 64-bit hash value only [109], making the use of shape patterns highly efficient.

Especially with the use of whole board patterns, the total number of possible shape patterns becomes even larger than the number of legal Go positions. To reduce the set of patterns to a reasonable number, frequently appearing shape patterns are typically *harvested* in a seperate step from a collection of move decisions. Only shape patterns that occur a given number of times in the move decision collection are regarded in later training and prediction steps.

Feature Pattern

In addition to shape patterns, so called feature patterns, that represent non-shape properties of state-action pairs are used. Table 5.1 lists the feature patterns that are used for the experiments presented in this chapter. In this table there are two groups of feature patterns that cope with the distance between two moves. To measure this distance between two moves $x = (x_1, x_2)$ and $y = (y_1, y_2)$ we used the following measure: $\text{dist}(x, y) = |x_1 - y_1| + |x_2 - y_2| + \max(|x_1 - y_1|, |x_2 - y_2|)$. The same measure was also used in [28]. The features are grouped in a manner that at most one feature per group can be active at the same time.

5.2 Bayesian Move Prediction Systems

We will now present three Bayesian prediction systems that were presented in literature, namely those of Coulom [28], Stern, Herbrich and Graepel [96] and Weng and Lin [103] together with according modifications we made to them in order to allow for a fair comparison of those systems in the context of move prediction for the game of Go. This presentation should mainly serve as a brief summary of existing prediction systems rather than being a complete introduction to the various approaches.

The input for all presented systems is a number of game positions, each annotated with the single move that is regarded to be the best. Once a new position is seen, each move is characterized with a team of patterns. Hence, after this preprocessing step, the input for all presented algorithms is a number of competitions between pattern teams, where one of the teams is marked as the winning team. For the remainder of this chapter we define all variables related to the winning team to be indexed with 1.

A common element of all considered methods is the definition of a ranking model, resulting in a single function L that represents the likelihood of the observed data in the face of given pattern strength parameter values. Using Bayes-Theorem in

[3] The Simple-Ko rule forbids immediate recaptures of single stones (cf. Section 2.1.1).

Table 5.1: List of feature patterns

#	Feature Pattern	Group
1	Capture of an opponent group that is adjacent to one of the player's groups that has only one liberty left	Capture
2	Recapture, i.e., a capture of an opponent group that was created by capturing one of the player's groups in the last turn	
3	Capture at an intersection where the opponent could play otherwise to connect to another group	
4	Other kinds of captures	
5	Extension, i.e., we extend a single liberty group	Extension
6	Self-atari, i.e., this move creates a group that could immediately be captured by the opponent	Self-atari
7	Atari (i.e. we will make an opponent group to remain with a single liberty only) while there is a Simple-Ko[3]on the board	Atari
8	Other kinds of Atari	
9	Distance 1 to the edge of the board	Edge
10	Distance 2 to the edge of the board	
11	Distance 3 to the edge of the board	
12	Distance 4 to the edge of the board	
13	Distance ≥5 to the edge of the board	
14	Distance 2 to the previous move	Previous move
15	Distance 3 to the previous move	
⋮		
29	Distance 17 to the previous move	
30	Distance >17 to the previous move	
31	Distance 2 to the move before the previous move	Before previous move
⋮		
46	Distance 17 to the move before the previous move	
47	Distance >17 to the move before the previous move	

the one way or another, this function is used to update the patterns strength parameters with the objective to maximize L for the training data.

5.2.1 Minorization-Maximization

In [28], Coulom proposed a whole-history rating concept that iterates several times over the whole training data with the objective to find the pattern strength values that maximize the likelihood of the given move decisions from the training set according to the generalized Bradley-Terry model (cf. Equation 5.1 in Section 5.1.2). Following the work of Hunter [54], Coulom iteratively applied the concept of Minorization-Maximization (MM), a hill climbing technique, on

$$L(\boldsymbol{\theta}) = \prod_{j=1}^{N} P(D_j|\boldsymbol{\theta}), \tag{5.4}$$

where N is the number of move decisions in the training set and $P(D_j|\boldsymbol{\theta})$ is the generalized Bradley-Terry model. By using the generalized Bradley-Terry model, Coulom benefits from the ability to abstract each move with a team of patterns and thus, to use in addition to shape patterns a larger number of additional feature patterns like, e.g., the distance to the move played just before. Note that L can be written as a function of each of the single strength parameters $\gamma_i = e^{\theta_i}$ allowing for partial optimization in one dimension.

As illustrated in Figure 5.2, MM is an optimization algorithm where, starting from an initial guess γ_0, a *minorizing* function $m(\cdot)$ for $L(\cdot)$ is build in γ_0 (i.e. $m(\gamma_0) = L(\gamma_0)$ and $\forall \gamma : m(\gamma) \leq L(\gamma)$) so that its maximum can be given in closed form. Finally the maximum γ_1 of m is chosen as an improvement over γ_0. Note that only in this paragraph, subscripts denote subsequent iterations of MM instead of indices into the vector of pattern strength values $\boldsymbol{\theta}$.

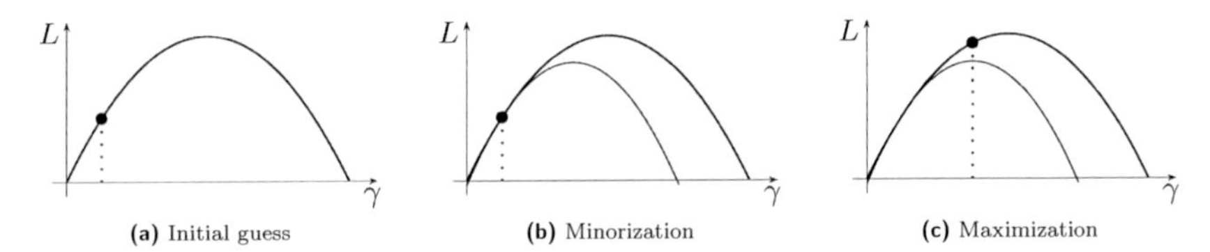

(a) Initial guess (b) Minorization (c) Maximization

Figure 5.2: Minorization-Maximization (Source: [28]).

In order to find the pattern strength parameter values that maximize function L as given in Equation 5.4, MM requires us to iterate a number of times over the complete training set to update the strength values $\gamma_i = e^{\theta_i}$ of all patterns until convergence. Given a number of competitions, i.e., move decisions, Coulom derived the following update formula for the individual pattern strength parameters:

$$\gamma_i \leftarrow \frac{W_i}{\sum_{j=1}^{N} \frac{C_{ij}}{E_j}}, \tag{5.5}$$

where N is the number of competitions in the training set, W_i is the number of times pattern i was a member of the winning team in all N competitions, C_{ij} is the strength of the team-mates of pattern i in competition j and E_j is the sum of the strength of all teams competing in j. For more details on the derivation of the update Formula 5.5 we refer to Coulom's paper [28]. We applied this approach following an implementation example published by Coulom on his homepage[4]. Here, after a first update with Formula 5.5 for all gamma values, he selects the feature group that lead to the biggest improvement towards minimizing L during its last update also for the next update, until the last change of the logarithm of L was smaller than some threshold for each feature group. In this comparison we used a threshold of 0.001.

5.2.2 Bayesian Ranking Model

The probability model for game decisions in this approach developed by Stern et al. [96] is based on the TM model. In addition to the pattern strength that is modeled as a Gaussian, they account for varying performances or playing styles of players that produced the training set by adding a fixed variance β^2 to a pattern's strength distribution. Hence, the performance of a pattern i with strength $\theta_i = \mathcal{N}(\mu_i, \sigma_i^2)$ is given by $x_i = \mathcal{N}(\theta_i, \beta^2)$. They characterize each state-action with a single combined shape-feature pattern, i.e., a shape pattern that additionally contains a tiny binary feature vector. Individual strength values are learned for each such combined pattern. Hence, the total number of patterns used by the system grows exponentially in the number of binary features considered.

Given a board position with k possible moves and corresponding strength values $\theta_1, \dots, \theta_k$, they give the following joint distribution for the probability that the pattern indexed with 1 is the best performing move:

$$p(i^* = 1, \boldsymbol{x}, \boldsymbol{\theta}) = \prod_{i=1}^{k} s_i(\theta_i) \prod_{j=1}^{k} g_j(x_j, \theta_j) \prod_{m=2}^{k} h_m(x_1, x_m), \tag{5.6}$$

[4] Rémi Coulom's website on MM is located at http://remi.coulom.free.fr/Amsterdam2007/.

where

$$\begin{aligned} s_i &= \mathcal{N}(\theta_i; \mu_i, \sigma_i^2), \\ g_j &= \mathcal{N}(x_j; \theta_j, \beta^2) \text{ and} \\ h_m &= \mathbb{I}(x_1 > x_m). \end{aligned}$$

Here, $\mathbb{I}(cond.)$ is the indicator function that equals 1 in case the given condition holds and 0 otherwise. The distribution is a product of small factors and can be represented as a graphical model, the so called factor graph. Figure 5.3 shows an example of such a graph. For tree like factor graphs, there exist message passing algorithms that allow for efficient Bayesian inference in the graph by exploiting the graphical structure of the underlying distribution[5]. Except of the h_m, all factors yield Gaussian densities. In order to keep computations simple, emerging non-Gaussian distributions are approximated with Gaussians. The resulting inference algorithm is called Expectation Propagation [75].

As a result, we can infer new values for the multivariate pattern strength distribution $\boldsymbol{\theta}$ by incorporating observations from a move decision. This new values can now act as the prior for incorporating the next observation. This makes the method a filtering algorithm also called *assumed density filtering*, where *assumed* stems from the approximation of emerging non-Gaussian distributions mentioned above.

The probability model used for inference in the Bayesian Full Ranking method of Stern et al. [96] is given by Equation 5.6. This distribution can be represented with a factor graph where each variable and each factor is represented by a node. The edges connect each factor node with the nodes of those variables the factor depends on. Stern et al. proposed a model that abstracts each move with exactly one shape-pattern with some binary non-shape properties encoded in it. We extended the model by introducing pattern teams as they were used in Coulom [28] by adding another level to the graph. This allows for the use of more non-shape patterns and makes a head-to-head comparison with Couloms approach possible. As some patterns can be member of more than one team, the graph might contain several cycles. Figure 5.3 shows the factor graph of this model with the extension to pattern teams. In this figure, θ_i now represents a team of patterns instead of a single pattern and θ_{ij} a member of team i. In order to cope with the introduced cycles we used so called *loopy-belief propagation* for message passing. The possibility to do this was already mentioned by Stern himself in [95]. The message passing algorithm is rather involved and we refer to [95] for a description of message passing in this particular setting with a derivation of all the update equations.

[5] See [72] for more information on message passing in factor graphs.

In the remainder of this chapter, we call the modified, pattern teams related approach *Loopy-Bayesian Ranking*.

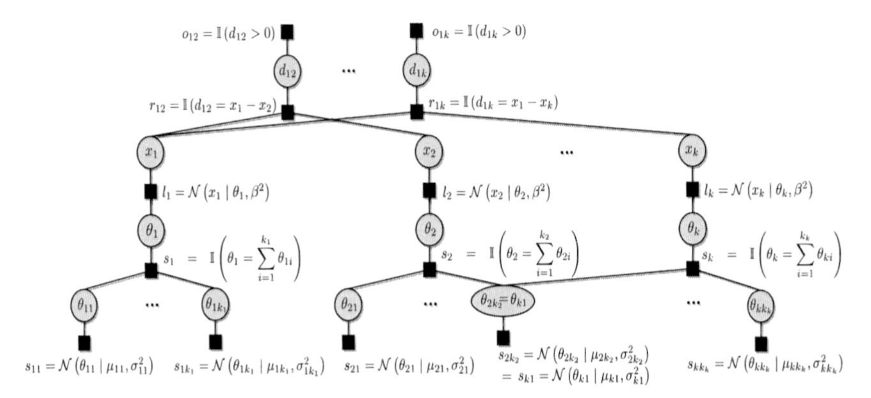

Figure 5.3: Factor graph of Bayesian Full Ranking Model (from [95]) extended to pattern teams.

5.2.3 Bayesian Approximation Ranking

Weng and Lin [103] present a more general Bayesian approximation framework for online ranking of players from games with multiple teams and multiple players and compare their method to the online ranking system TrueSkill™ [49] that was developed and is used in commercial products by Microsoft. TrueSkill has a lot of similarities to the Bayesian Full Ranking Model described above. The framework presented in [103] can be used with the Bradley-Terry as well as the Thurstone-Mosteller and other ranking models. Like the Bayesian Full Ranking system of Stern et al., the players' strength values θ_i are updated after each new observation and are assumed to be normal distributed with mean μ_i and variance σ_i^2, i.e., $\theta_i = \mathcal{N}(\mu_i, \sigma_i^2)$. As a main achievement, Weng and Lin were able to construct computationally light-weight approximated Bayesian update formulas for the mean and variances of the θ_i by approximating the integrals that naturally show up in Bayesian inference analytically.

In this comparison we applied the algorithm based on the Bradley-Terry model to the game of Go in order to compare it to the approaches of Stern et al. and Coulom.

As for the Bayesian Full Ranking model, Weng and Lin use a probability model for the outcome of a competition between k teams that can be written as a product of factors that each involve only a small number of variables:

$$L(\boldsymbol{\theta}) = \prod_{i=1}^{k} \prod_{q=i+1}^{k} P(\text{outcome between team } i \text{ and team } q|\boldsymbol{\theta}),$$

where P is given by the Bradley-Terry model.

For their system they consider a complete ranking of the teams whereas in case of a move decision in Go we have exactly one winning team of patterns and all the others are ranked second. Hence, in addition to the aforementioned function L, we made experiments with a simplified function

$$L_{\text{Go}}(\boldsymbol{\theta}) = \prod_{i=2}^{k} P(\text{team 1 beats team } i|\boldsymbol{\theta}).$$

Recall that we defined the winning team's index to be 1. In all our preliminary experiments we achieved significantly better prediction rates using L_{Go} instead of L. Hence, in section 5.3 all experiments were made using L_{Go}. In a comment on our paper [104], Weng pointed out, that the use of L_{Go} might introduce some undesired bias in the parameter updates, as the variance of the winning team gets reduced much faster than for the losing teams. Hence, further improvements might be possible by accounting for this.

The actual complete algorithm for the update of the pattern strength parameter vectors $\boldsymbol{\mu}, \boldsymbol{\sigma^2}$ on the observation of a new competition result towards maximizing L_{Go} is shown in Algorithm 1. These slightly modified algorithm is termed *Bayesian Approximation Ranking* in the remainder of this chapter.

5.2.4 Probabilistic Ranking

In addition to the three above mentioned Bayesian prediction systems, we add a very simple system to the comparison that we named *Probabilisitc Ranking*. To train this model on a number of move decisions, we simply count for each single pattern the number of times it was seen in the training set, as well as the number of times the pattern was in a winning move's team of patterns. The pattern strength values of each pattern θ_i are then computed by:

$$\theta_i = \frac{\text{number of times pattern } i \text{ was in a winning move's team}}{\text{number of times pattern } i \text{ was seen in the training set}} .$$

For prediction, the value of a move represented by a set of patterns F is computed by $\sum_{j \in F} \theta_j$. The moves are then ranked according to their value, where the highest valued move is ranked first.

Algorithm 1: Bayesian Approximation Ranking.

input : $(\mu_1, \sigma_1^2), \ldots, (\mu_k, \sigma_k^2)$, with (μ_1, σ_1^2) winning
output: Updated values $\boldsymbol{\mu}$ and $\boldsymbol{\sigma^2}$
$\beta^2 := 13$

for $i = 1, \ldots, k$ **do**
  $\mu_i = \sum_{j=1}^{n_i} \mu_{ij}$
  $\sigma_i^2 = \sum_{j=1}^{n_i} \sigma_{ij}^2$
end

/* Update team strength */

for $i = 1, \ldots, k$ **do**
  for $q = 1, \ldots, k, \mathit{rank}(i) \neq \mathit{rank}(q)$ **do**
    $c_{iq} = \sqrt{\sigma_i^2 + \sigma_q^2 + 2\beta^2}$
    $\hat{p}_{iq} = \frac{e^{\mu_i/c_{iq}}}{e^{\mu_i/c_{iq}} + e^{\mu_q/c_{iq}}}$
    if $i = 1$ **then**
      $\Omega_1 \leftarrow \Omega_1 + \sigma_1^2 \frac{\hat{p}_{q1}}{c_{1q}}$
      $\Delta_1 \leftarrow \Delta_1 + \frac{\sigma_1}{c_{1q}} \frac{\sigma_1^2}{c_{1q}^2} \hat{p}_{1q} \hat{p}_{q1}$
    end
  end
  if $i > 1$ **then**
    $\Omega_i \leftarrow -\sigma_i^2 \frac{\hat{p}_{i1}}{c_{i1}}$
    $\Delta_i \leftarrow \frac{\sigma_i}{c_{i1}} \frac{\sigma_i^2}{c_{i1}^2} \hat{p}_{i1} \hat{p}_{1i}$
  end

/* Update pattern strength */

  for $j = 1, \ldots, k_i$ **do**
    $\mu_{ij} \leftarrow \mu_{ij} + \frac{\sigma_{ij}^2}{\sigma_i^2} \Omega_i$
    $\sigma_{ij}^2 \leftarrow \sigma_{ij}^2 \cdot \max\left\{1 - \frac{\sigma_{ij}^2}{\sigma_i^2} \Delta_i, \epsilon\right\}$
  end
end

5.3 Experimental Results

We made a series of experiments with the four algorithms presented in section 5.2. The main focus was on the comparison of the core learning routines of all approaches under equal conditions, with respect to the achievable prediction performance as well as the time needed to learn model parameters. For all approaches, the procedure to produce predictions, once the model parameters were trained, is rather equal and simple, so we did not investigate computational requirements for this step. We will first give the experimental setup to show and discuss the results afterwards.

5.3.1 Setup

All experiment computations were performed on a single core of an Intel X5650 CPU running at 2.67 GHz. The machine was equipped with 36 GB main memory. The prediction systems were trained with up to 20,000 records of Go games played on the KGS-Go-Server[6] by strong amateur players on the 19×19 board size. The games were obtained from u-go.net[7]. The test set contained 1,000 game records from the same source, however, the test set was disjoint from the training set. Apart from this, a subset of the feature patterns presented in Coulom [28] was used.

5.3.2 Results

Harvesting of Shape Patterns

The harvesting of shape patterns on a training set of game records has to be done a single time only. The result is then used for all further experiments with the distinct prediction systems. It is necessary to decide on a small subset of all possible shape patterns that is regarded to contain most of the shape patterns that are relevant in actual game play. We performed a single harvesting run for each of three differently sized training sets (5,000, 10,000 and 20,000 game records) to produce the sets of shape patterns that were considered by the learning systems afterwards. These sets were built from all shape patterns that were seen at least 10 times during harvesting, considering only patterns around moves that were actually played in the records.

We give some insights into the kind of shape patterns that were harvested on the training set of 10,000 game records. Figure 5.4 shows the percentage of occurrences

[6]The website of the KGS-Go-Server is located at `http://www.gokgs.com/`

[7]A large database of Go game records is available online at `http://u-go.net/gamerecords/`

of patterns of different sizes at different phases of the games. A phase lasts for 30 moves. One can see that, in early game phases with only a few pieces on the board, mostly large patterns are matched. Naturally, as the positions get more complex because of more pieces are placed on the board, mostly small patterns are matched more than 10 times in later game stages.

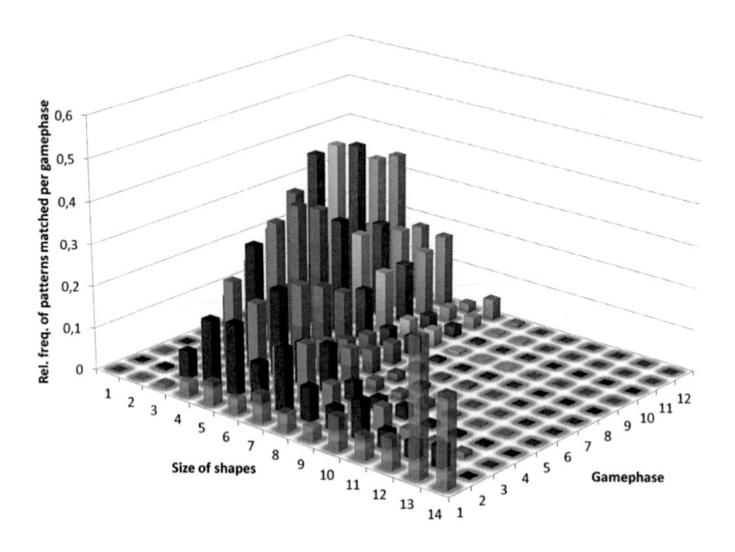

Figure 5.4: Relative frequency of encountered shape pattern sizes per game phases in 10,000 games.

Empirical Comparison of the Prediction Systems

The result of the harvesting step described above is a set of relevant shape-patterns that are likely to be regularly encountered in Go games. Together with the feature patterns listed in Section 5.1.4, we obtain the fixed set of patterns that we used for further experiments with the different prediction systems introduced above. Here we first present results of experiments regarding the achieved prediction quality of the different systems. We then look at the computational overheads induced by training the models and conclude with some experiments targeting at directions for further improvements of move prediction systems.

Figure 5.5 shows the cumulative distribution of finding the expert move within the first n ranked moves for the four prediction systems with a training set size of 10,000 games. The rate of ranking the expert move first is, for each system, larger as in the experimental results presented in the corresponding original papers. This might be the case because of a larger training set for MM and the use of pattern teams for the Loopy-Bayesian Ranking system. Even further, the curves are pretty similar for all systems except for the very simple probabilistic ranking.

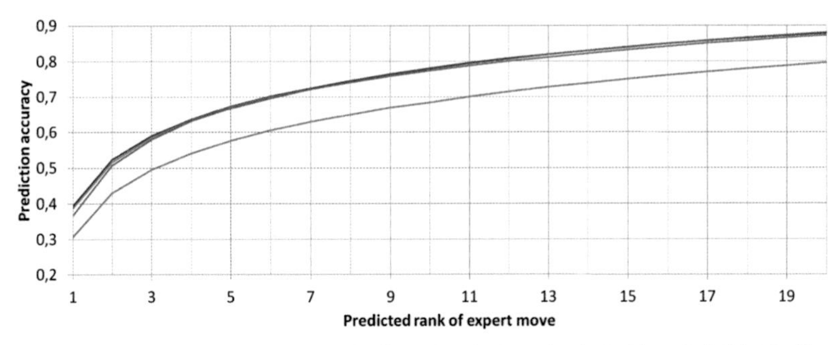

Figure 5.5: Cumulative distribution of finding the expert move within the first n ranked moves.

Figure 5.6 gives a closer look to the differences between the Bayesian systems by showing the distance of the cumulative distributions to the one of the probabilistic ranking system. Here we can see a small but significant better performance of the Loopy-Bayesian Ranking system and MM over Bayesian Approximation Ranking in ranking the expert move first. However, for finding the expert move in the first n ranked moves for $n \geq 3$ all systems perform almost equal.

Table 5.2 shows the development of prediction rates for different training set sizes. It can be seen that all systems benefit from increasing training set sizes. We did not conduct experiments with MM for 20,000 training games due to memory limitations.

Table 5.2: Prediction rates for different training set sizes

	5,000	10,000	20,000
Minorization-Maximization	37.00%	37.86%	-
Loopy Bayesian Ranking	36.36%	37.35%	38.04%
Bayesian Approximation Ranking	34.24%	35.33%	36.19%
Probabilistic Ranking	29.16%	30.17%	30.92%

Regarding the distributions given in Figures 5.5 and 5.6, we should note that there are a lot more moves available in early game stages than in the end games, when the board is almost full of pieces. Therefore we measured the rank errors of the systems at different game phases. The rank error is the fraction of the expert move rank assigned by the system over the number of available moves. Figure 5.7

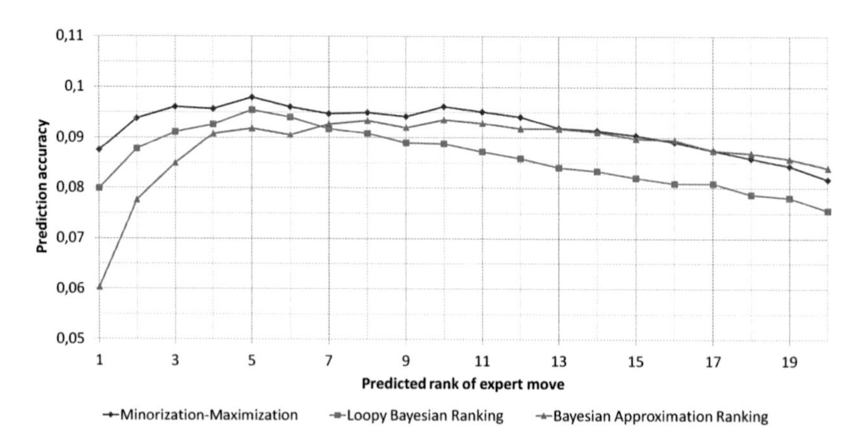

Figure 5.6: Difference of cumulative prediction rank distribution to probabilistic ranking distribution.

shows these ranking errors in a box plot. The boxes range from the lower to the upper quartile of the data and contain the median line. The whiskers cover 95% of the data. Data points outside the whiskers range are explicitly shown as proper data points.

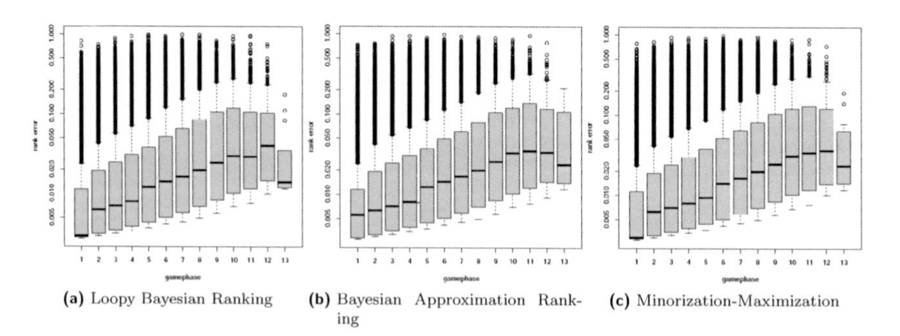

(a) Loopy Bayesian Ranking **(b)** Bayesian Approximation Ranking **(c)** Minorization-Maximization

Figure 5.7: Ranking errors at different stages of the game

Figure 5.8 shows the average time needed by the systems per game during parameter training. Not surprisingly, Loopy Bayesian Ranking required a lot more time than Bayesian Approximation Ranking. MM was iterated until convergence to a certain level, leading to varying time consumption for different training set sizes.

A further interesting point is the dependency of the prediction rate of expert moves on the size of the shape pattern that was matched for the particular move. The

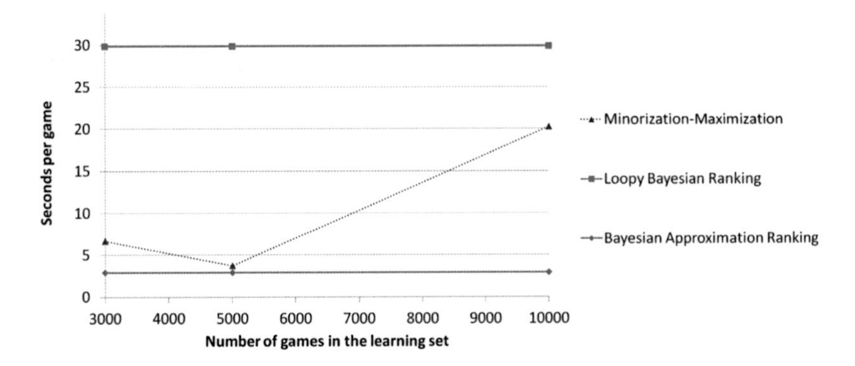

Figure 5.8: Time needed per game on different training set sizes

larger the pattern, the more information about the actual board configuration is encoded in the pattern. So we expect to get better prediction rates, the larger the matched patterns are. Figure 5.9 shows the achieved average prediction rates for the different pattern sizes. The curve in the background of the figure shows the distribution over all matched patterns of different sizes.

It can be seen that most of the patterns that are matched during a game are of size 3 to 5. Of those, pattern sizes of 3 and 4 show the significantly worst prediction rates. In order to improve here, we added a small feature vector to shape patterns of size 2 to 4. Here we took exactly the same features as done by Stern et al. [96]. Figure 5.10 shows the achieved improvements for the different ranking systems at different game phases. Following Figure 5.4, small shape patterns are mostly matched in later game phases, which explains the increasing impact of this modifications for later game phases.

Inspired by our results, recently Wistuba and Schmidt-Thieme [105] realized an application of factorization machines to move prediction in computer Go that allows for efficient learning of sparse first order pattern dependencies. With their prediction system, called *Latent Factor Ranking* (LFR), they obtained remarkably high prediction rates. Figure 5.11 shows their results in form of the prediction accuracy obtained at different game phases. The abbreviations MM and BAR in the ledgend of their diagram denote Minorization-Maximization and Bayesian Approximation Ranking, respectively. The setup of the experiments was identical to ours. LFR1 and LFR5 are different versions of their developed prediction system Latent Factor Ranking.

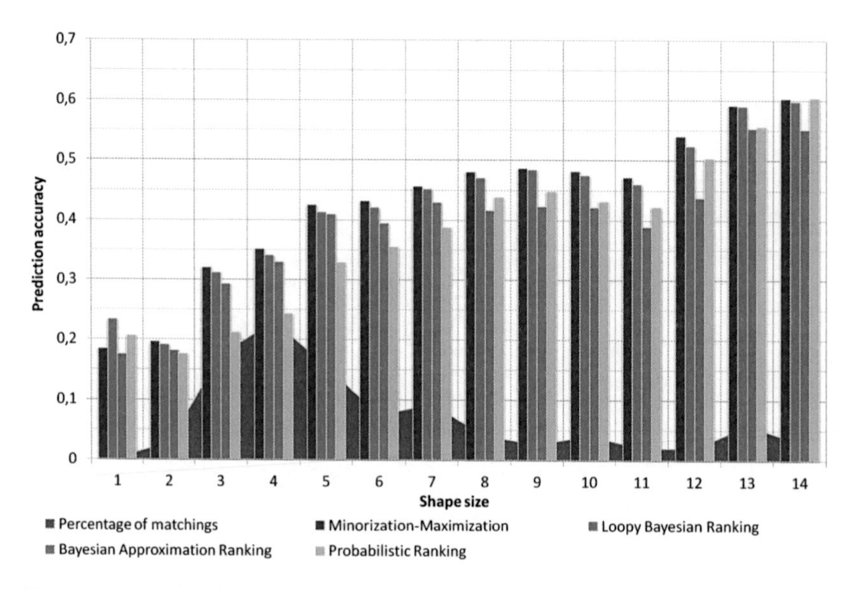

Figure 5.9: Average prediction rates in relation to the size of the pattern that was matched at the expert move.

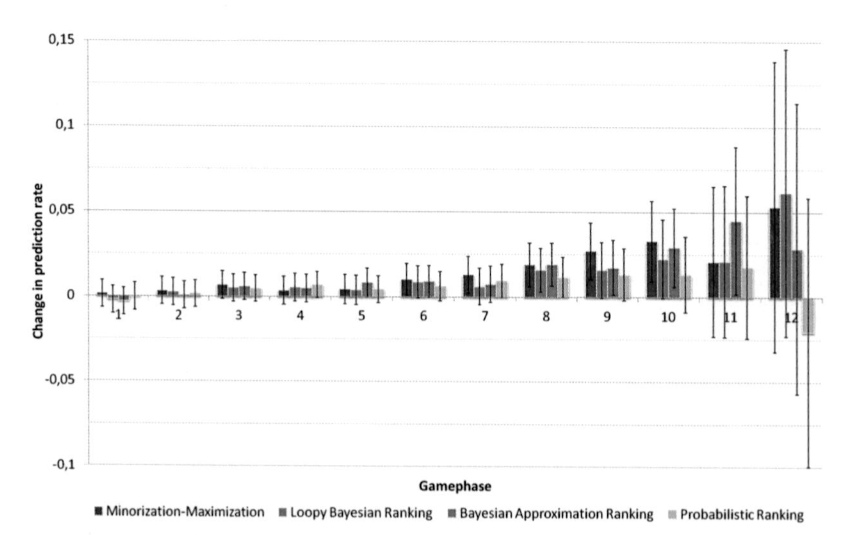

Figure 5.10: Relative strength improvements from adding a small feature vector to shape patterns of size 2 to 4 at different game phases.

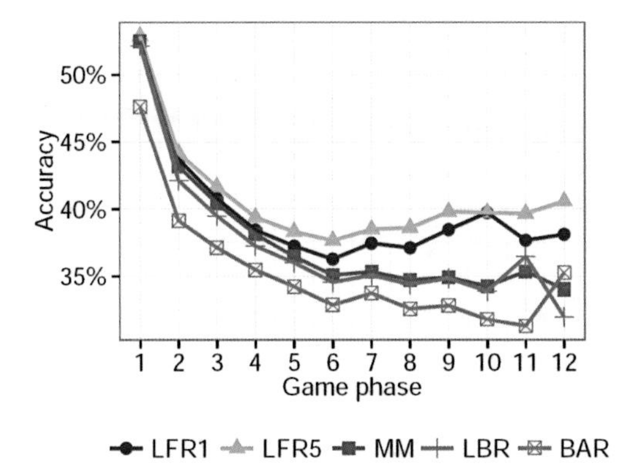

Figure 5.11: Move prediction accuracy in different game phases. Each game phase consists of 30 turns. (Source: [105])

5.4 Chapter Conclusion

In this chapter we compared three Bayesian systems for move prediction under fair conditions and investigated their performance by using them for move prediction with the game of Go. We observed that all prediction systems have a comparable performance concerning their prediction accuracy but differ significantly in the amount of time needed for model parameter training. Additionally we presented some insight into the use of shape patterns that are prominent for modeling in the game of Go. Following our observation, we were able to improve our pattern system further and gained some improvement in terms of prediction rate. Partly inspired by our findings, a novel move prediction system was developed for Computer Go by Wistuba et al. [105].

The outcome of the investigations presented in this chapter might serve as a base for further improving the performance of state-of-the-art MCTS based Go programs by online adaptation of Monte-Carlo playouts. Recent strength improvement of Go playing programs, especially on large board sizes, were made by adding more and more domain knowledge to the playout policies that made them smarter and capable to *understand* more difficult situations that require selective and well focused play. In order to further improve these policies we, among others (e.g., [33][7][8]), try to adapt simulation policies to the actual board position using insights gathered by MC sampling. As a conclusion from the results presented in this chapter, Bayesian Approximation Ranking appears to be a promising candidate for the use with online model adaptation in such time critical environments. But also the recently developed Latent Factor Ranking algorithm exposes promising properties that deserve further investigations.

CHAPTER 6

MCTS Driven Position Analysis[1]

A frequently mentioned limitation of Monte-Carlo Tree Search (MCTS) based Go programs is their inability to recognize and adequately handle capturing races, also known as *semeai*, especially when several of them appear simultaneously. This essentially stems from the fact that certain group status evaluations require deep lines of correct tactical play that somewhat oppose to the exploratory nature of MCTS. In this chapter we present a technique for heuristically detecting and analyzing regions in Go positions that are subject to ongoing fights and accordingly contain groups of uncertain status. This is done during the search process of a state-of-the-art MCTS implementation. We evaluate the strength of our approach on game positions that are known to be difficult to handle even by the strongest Go programs to date. Our results show a clear identification of semeais that are contained in this positions and thereby advocate our approach as a promising heuristic for the design of future MCTS simulation policies.

6.1 Motivation

A crucial part of MCTS algorithms is the playout policy used for move decisions in game states where statistics are not yet available (cf. Section 3.4). In general, the playout policy drives most of the move decisions during each simulation. Moreover, in MCTS, the simulations' terminal positions are the main source of data for all

[1]We presented parts of the content of this chapter before at the International Conference on Computers and Games 2013 (cf. [46]). The work was inspired by Ingo Althoefer, who observed a correlation between the existence of multiple peaks in score histograms produced during MCTS searches and misevaluations of strong MCTS based Go programs. He pointed the author of this thesis to this apparent dependency at the European Go Congress in August 2012 and hereby laid the foundation of this work.

remaining computations. Accordingly, for the game of Go there exists a large number of publications about the design of such policies, e.g., [43][27][28]. One of the objectives playout designers pursue focuses on balancing simulations to prevent biased evaluations [43][93][51]. Simulation balancing targets at ensuring that the policy generates moves of equal quality for both players in any situation. Hence, adding domain knowledge to the playout policy for attacking also necessitates adding domain knowledge for according defense moves. One of the greatest early improvements in Monte-Carlo Go were sequence-like playout policies [43] that highly concentrate on local answer moves. They lead to a very selective search. Further concentration on local attack and defense moves improved the handling of some tactical fights and hereby contributed to additional strength gain of MCTS programs. However, adding more and more specific domain knowledge with the result of increasingly selective playouts, we open the door for more imbalance. This in turn allows for severe false estimates of position values. Accordingly, the correct evaluation of, e.g., semeai is still considered to be extremely challenging for MCTS based Go programs [81]. This holds true, especially when they require long sequences of correct play by either player. In order to face this issue, we search for a way to make MCTS become aware of probably biased evaluations due to the existence of semeai or groups with uncertain status. In this chapter we present our results about the analysis of score histograms to infer information about the presence of groups with uncertain status. We heuristically locate fights on the Go board and estimated their corresponding relevance for winning the game. The developed heuristic is not yet used by the MCTS search. Accordingly, we cannot definitely specify and empirically prove the benefit of the proposed heuristic in terms of playing strength. We further conducted experiments with our MCTS Computer Go engine Gomorra on a number of 9×9 game positions that are known to be difficult to handle by state-of-the-art Go programs. All these positions include two ongoing capturing fights that were successfully recognized and localized by Gomorra using the method presented in the remainder of this chapter.

6.2 Background and Related Work

At the end of each MC simulation we obtain a terminal position and a corresponding score value. The clustering of MC simulations into groups of similar scores, takes a central role in our proposed method for identifying local fights in game positions. The distribution of scores obtained at the single simulations' terminal positions appears to be almost normally distributed when performing an MCTS search on an even position with no tactical fights present. Figure 6.1 shows the normalized histogram of the scores obtained from 128,000 MCTS simulations on

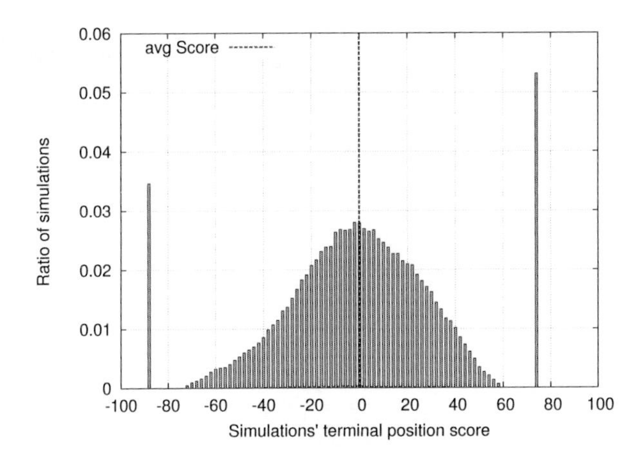

Figure 6.1: Normalized histogram of scores obtained from 128,000 MCTS simulations on an empty 9×9 Go board with komi 7.

the initial empty 9×9 Go position[2]. The dotted line marks the average score of all simulations.

Looking at the histogram, we make two additional observations. At first, there is a spacing between each bucket. This is the case, because the score difference between distinct terminal positions is most often even. To obtain two positions with an odd score difference it needs for rather rarely appearing seki (cf. Section 2.1.1, page 2.1.1). Second, the two extreme scores -88 and 74, that correspond to the terminal positions with only white stones resp. only black stones remaining on the board, appear quite often on the 9×9 board. Note that the tiniest possible living group, i.e., the tiniest group that contains two eyes, already has a size of 8, explaining the histogram values of zero next to the most extreme scores.

Let us consider the more complex situation depicted in Figure 6.2a, that was designed by Shih-Chieh (Aja) Huang (6 Dan) and Martin Müller as part of a regression test suite for their Go playing program FUEGO. The position contains two semeais, one on the upper side of the board, the other on the lower right. In Figure 6.2b, the two groups that belong to the upper side semeai are marked by circles. When the game continues, only one of these two groups can survive. Due to a sufficient number of outer liberties, marked with the letter o in the figure,

[2]The komi (cf. Section 2.1.1, page 10) was set to 7

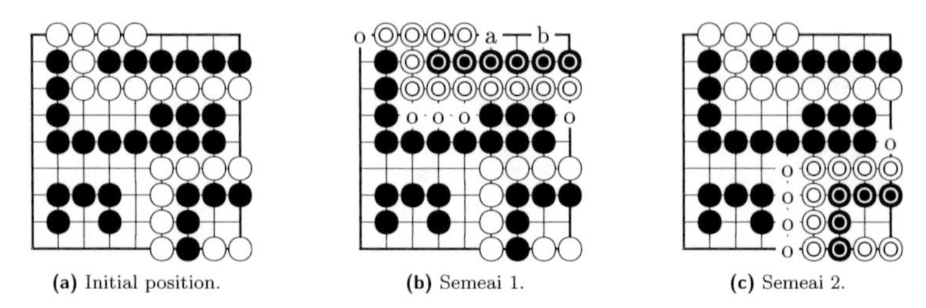

(a) Initial position. (b) Semeai 1. (c) Semeai 2.

Figure 6.2: Go position with two semeai. It is Black to move. Black should realize that both semeai will be won by White.

it only needs an average skilled player to realize that White will win this semeai. Black would need to place a stone on both of the intersections marked with a and b in the figure, in order to create 2 eyes. If White plays on one of these intersections first, Black must resort to capturing the whole white group but will be short of liberties and accordingly die first. Figure 6.2c shows the corresponding groups involved in the semeai at the lower right of the board. Also this second semeai will clearly be won by the white player.

In such complex positions, we might encounter evaluation uncertainties from MC-simulations regarding the survival of certain groups of stones. This is the case because the semi-random playout policies are in general not capable of playing out lengthy tactical fights correctly. The uncertainties will likely result in multi-modal score distributions. Figure 6.3 shows the histogram obtained from 128,000 MCTS simulations computed for the Go position discussed above, that contains two ongoing capturing races, i.e., semeai. We can distinguish four major score clusters in the intervals $[-20, 0], [15, 30], [30, 40]$ and $[74, 74]$. A fifth small cluster can be discovered in the score interval $[0, 10]$. This small cluster will be ignored by our algorithm due to its tiny proportional value. A distinction of four major clusters is in line with our expectation when looking at a position with two semeais. Each semeai will be won by either Black or White, resulting in a total of four possible outcomes. In case each outcome yields a different score, we expect to see four score clusters in the histogram.

Note that MCTS in its original form is not informed about the multi modal distribution of outcomes but guides simulations mostly based on the simulations' mean outcome. In Figure 6.3, the dotted blue line shows the mean outcome to be about 30 points, hence a solid win for Black. But in fact, the leftmost cluster, representing a win for White, contains the correct position evaluation.

In the remainder of this chapter we present how to computationally obtain the information about the existence of location of score cluster in histograms that were

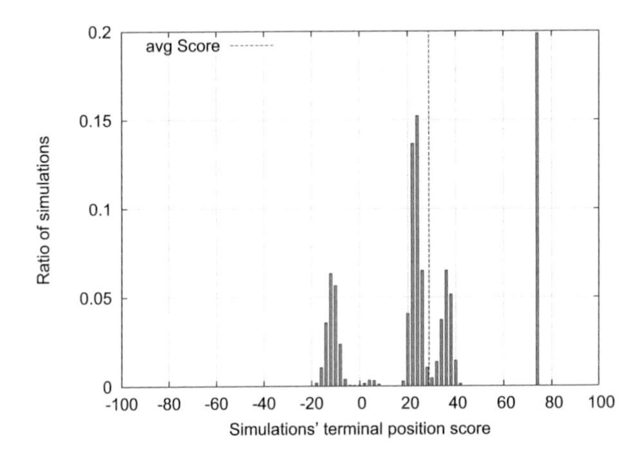

Figure 6.3: Normalized histogram of scores obtained from 128,000 MCTS simulations on a 9×9 Go position with two semeai.

generated during MCTS searches. We furthermore present a method for localizing the respective groups with uncertain state that lead to each of the score clusters on the Go board.

We decided to use a mean shift algorithm for detecting clusters in score histograms. Mean shift is a simple mode-seeking algorithm that was initially presented in [39] and more deeply analyzed in [26]. It is essentially a gradient ascent procedure on an estimated density function that can be generated by Kernel Density Estimation on sampled data. Modes are computed by iteratively shifting the sample points to the weighted mean of their local neighboring points until convergence. Weighting and locality are hereby controlled by a kernel and a bandwidth parameter. We obtain a clustering by assigning each point in the sample space the nearest mode computed by the mean shift algorithm. We refer to Section 6.3.2 for explicit formulae and detailed information about our implementation.

The stochastic mapping of clusters to relevant regions on the Go board is realized by the use of MC-heuristics [14][28]. Starting in 2009 several researchers came up with the idea of using an intuitive covariance measure between controlling a certain point of a Go board and winning the game. The covariance measure is derived from the observation of MC simulation's terminal positions[29][3][79][8].

[3] Available online: `http://remi.coulom.free.fr/Criticality/`.

Such kind of measures are most often called *(MC-)Criticality* when being used in the context of Computer Go. The particular publications around this topic differ slightly in the exact definition and more remarkably in the way of integrating MC-Criticality with the general MCTS framework. Coulom [29] proposed the use of MC-Criticality as part of the feature space of a move prediction system that in turn is used for playout guidance in yet rarely sampled game states, while Pellegrine et al. [79] used the criticality measure with an additional term in the UCT formula. Baudis and Gailly [8] argued for a correlation of MC-Criticality and RAVE and consequently integrated both.

6.3 MC-Criticality Based Semeai Detection

In this section, we present our approach for detecting and localizing capturing races (jap.: semeai) in positions of the game of Go by the clustering of MC simulations according to their respective score and the computation of cluster-wise MC-criticality. When performing an MCTS search on a game position, a number of randomized game continuations called simulations are generated from the position under consideration. Each of these simulations ends in some terminal game position that can be scored according to the game rules. In case of Go, the achievable score per simulation ranges from about -361 to +361 when using Chinese scoring rules[4] (cf. Section 2.1.1, page 8). A common first step for obtaining information about the score distribution is the construction of a score histogram that can be interpreted as an empirical density function by appropriate scaling. Assuming that the presence of semeai likely results in more than one cluster of simulation scores (depending on whether the one or the other player wins) we are interested in identifying such clusters and the corresponding regions on the Go board that are responsible for each particular cluster. Accordingly, semeai detection with our approach is limited to cases in which different semeai outcomes lead to distinguishable final game scores.

In the following, we first introduce some notations and afterwards step through our method starting with clustering.

6.3.1 Notations

Let $S \subseteq \mathbb{Z}$ be the discrete set of achievable scores. Having built a histogram H of a total of n simulation outcomes, we write $H(s)$ for the number of simulations that achieved a final score of $s \in S$, hence $\sum_{s \in S} H(s) = n$. We denote the average

[4] The player that makes the second move in the game is typically awarded some fixed bonus points, called komi, to compensate for the advantage the other player has by making the first move. Typical komi values are 6.5, 7 and 7.5, depending on the board size. Accordingly, the score range might become asymmetric.

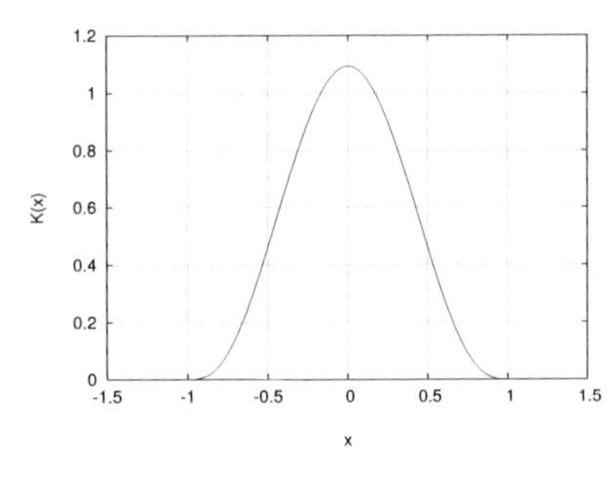

Figure 6.4: The triweight kernel $K(x)$

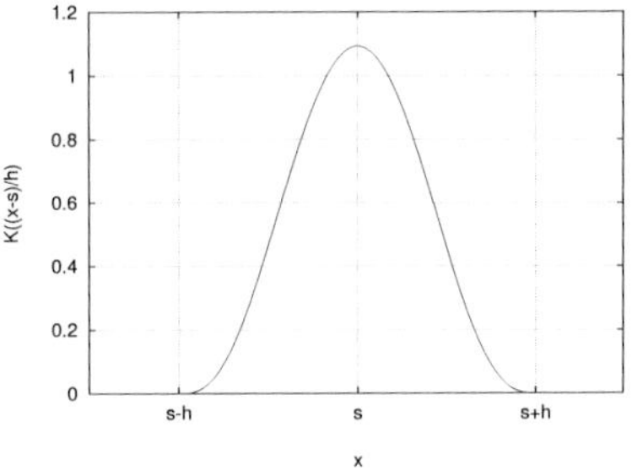

Figure 6.5: $K\left(\frac{x-s}{h}\right)$ as used in $\hat{f}(y)$

score of n simulations by $\overline{s} = \sum_{s \in S} H(s)/n$. Each element c of the set of score clusters C is itself an interval of scores, hence $c \subseteq S$. All clusters are disjunct in respect to the score sets they represent. We write $c(s)$ for the single cluster to which a score s is assigned.

6.3.2 Clustering

As mentioned in Section 6.2 we use a mean-shift algorithm for mode seeking and clustering in score histograms. The algorithm is based on a technique called Kernel Density Estimation (KDE), that allows for computing an estimated density function for histograms. As the name suggests, it makes use of a kernel. A kernel function $K : \mathbb{R} \rightarrow \mathbb{R}$ is a symmetric weighting function with the properties of being non-negative and real-valued integrable while additionally satisfying

$$\int_{-\inf}^{+\inf} K(x)dx = 1 \ .$$

In this chapter we solely use the so called triweight kernel K that is defined as

$$K(x) = \frac{35}{32}\left(1 - x^2\right)^3 \mathbb{I}(|x| \leq 1) \ ,$$

where $\mathbb{I}(cond.)$ denotes the indicator function that equals to 1 if the condition in the argument is true and to 0 otherwise. Figure 6.4 shows a plot of the kernel. As can be seen, the kernel has bounded support between -1 and 1. Note that $K^*(x) := \lambda K(\lambda x)$ for any kernel $K(x)$ also satisfies the above mentioned properties of a kernel if $\lambda > 0$. Hence, we can change the range of our kernel from $[-1, 1]$ to $[-h, h]$ by setting $\lambda = 1/h$. Figure 6.5 shows the curve of $K\left(\frac{x-s}{h}\right)$. In this context h is called the *bandwidth* parameter of the kernel.

Kernel Density Estimation (KDE)

When applying KDE to a normalized score histogram $H(s)/n$ like the one shown in Figure 6.3, we use the kernel to compute an estimated probability density value for each score s as the weighted mean of $H(s)/n$ and its local neighborhood. Here the weighting is given by the kernel function. Hence, when perceiving $f(s) = H(s)/n$ as an empiric density function, using KDE with the triweight kernel with bandwidth parameter h, we obtain a smooth estimated density function $\hat{f}$:

$$\hat{f}(y) = \frac{1}{nh} \sum_{s \in S} H(s) K\left(\frac{y-s}{h}\right) .$$

The size of the neighborhood considered in the estimation of each score's density value solely depends on the bandwidth parameter h. The bandwidth should be chosen in accordance to the certainty of the data that makes up the empiric density function, i.e., the histogram. When only few simulations were done and/or the simulations' scores are of high variance, a larger bandwidth parameter should be used to obtain a smooth estimate that ignores non-significant irregularities in the empiric density function. To determine an appropriate value of the bandwidth parameter h based on the number of simulations n and the observed variance of the simulations' scores $\hat{\sigma}$, we use *silverman's rule of thumb* [94] (p. 48):

$$h = C_{\text{K}} \cdot \hat{\sigma} n^{-\frac{1}{5}} \quad \text{with } \hat{\sigma}^2 = \frac{\sum_{s \in S}(s-\overline{s})^2 H(s)}{n-1} ,$$

with C_{K} being a constant factor that depends on the actual kernel used. Silverman gives the factor value of 3.15 to be used with the triweight kernel.

An example for a normalized histogram and the corresponding estimated density function $\hat{f}$ can be found in Figure 6.6, where $\hat{f}$ is represented by the black curve.

Mean-Shift

The mean-shift algorithm can be used to seek for modes of $\hat{f}$. Given some initial score $x \in S$ and the kernel function used for determining $\hat{f}$, the mean-shift algorithm computes the weighted mean m_x of all scores in the neighborhood of x that are covered by the kernel, i.e., $[x-h, x+h]$ when using a bandwidth parameter of h. Here, the weighting is according to the Kernel function. By thereafter setting x to m_x and repeating the procedure until convergence, the algorithm iteratively shifts x towards a mode of $\hat{f}$.

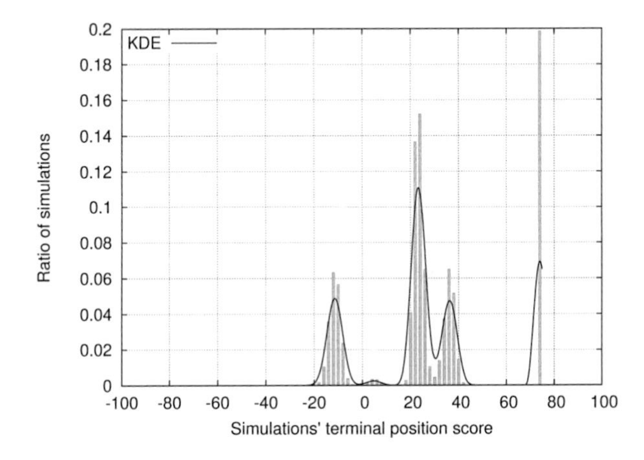

Figure 6.6: Estimated density function computed with KDE using the triweight Kernel.

To find all modes of $\hat{f}$ in practice, we initialize proper mode variables m_s for each score $s \in S$ to the respective score itself, i.e., $m_s = s$. The mean-shift algorithm now iteratively updates this mode variables by

$$m_s = \frac{\sum_{s' \in S} H(s')K\left(\frac{m_s - s'}{h}\right) s'}{\sum_{s' \in S} H(s')K\left(\frac{m_s - s'}{h}\right)}$$

until convergence. Figure 6.7a illustrates the initial positions of the single mode variables m_s when being initialized on our example histogram by crosses. Figures 6.7b to 6.7f then show their movement towards the actual modes of $\hat{f}$ during the first five mean-shift iterations. The different conversion points of the mode variables are the positions of the modes of $\hat{f}$. As in this example, a satisfying result is obtained already after 5 iterations, the procedure might need several hundred iterations in less extreme cases. For example, for the empty 9×9 board yielding a histogram as depicted in Figure 6.1 on page 101, mean-shift needs about 200 iterations until convergence.

Clustering

We finally build one score cluster for each mode and write m_c for the position of the mode that corresponds to cluster c. To account for estimation errors and sample variances we only consider clusters corresponding to mode positions m with

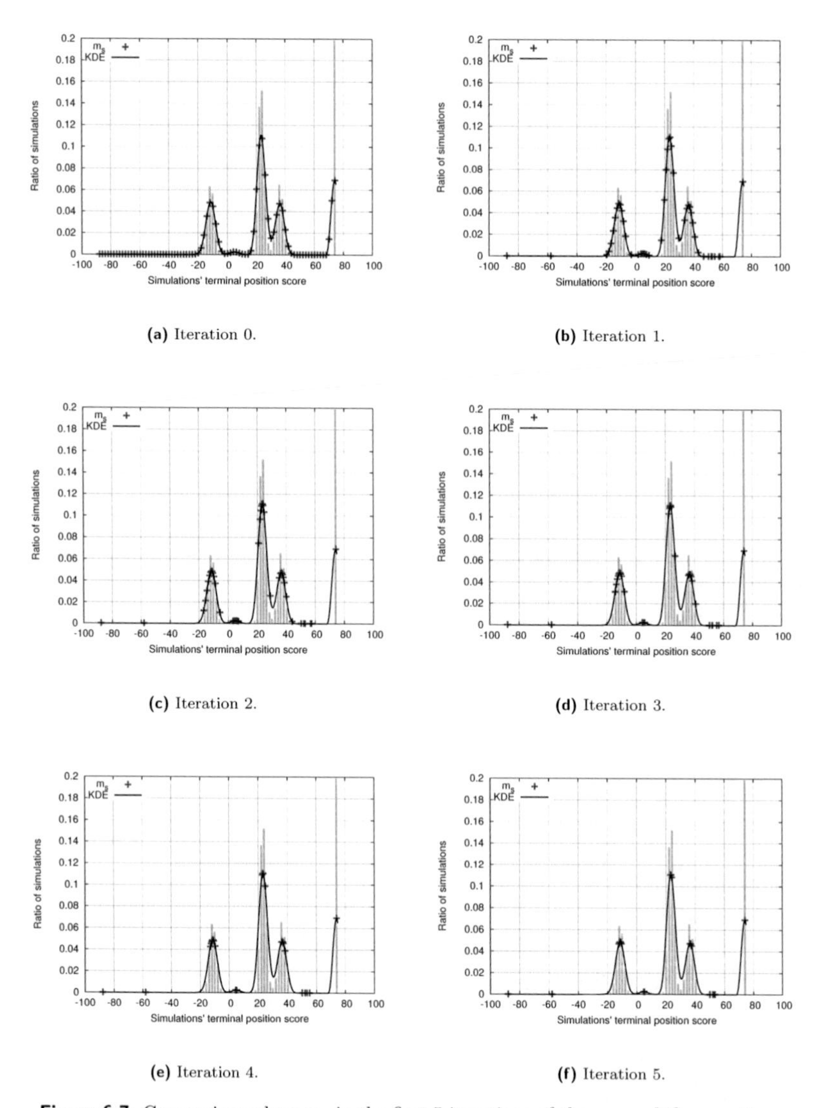

(a) Iteration 0.

(b) Iteration 1.

(c) Iteration 2.

(d) Iteration 3.

(e) Iteration 4.

(f) Iteration 5.

Figure 6.7: Converging values m_s in the first 5 iterations of the mean-shift operation.

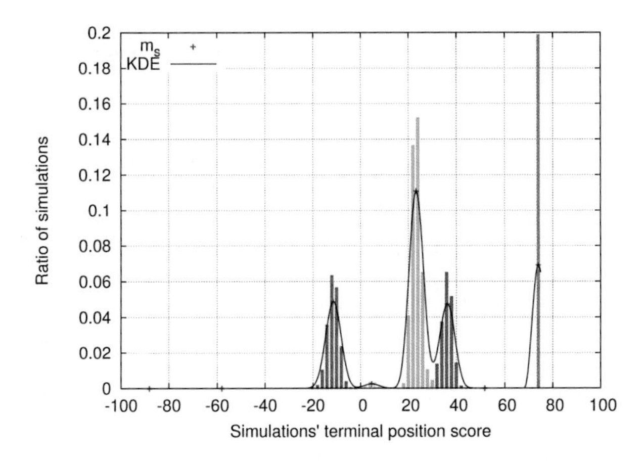

Figure 6.8: The four score clusters derived from means m_s with $\hat{f}(m_s) \geq 0.01$.

$\hat{f}(m) \geq T$ for some appropriate threshold T. Each score s is then assigned to the cluster $c(s) = \text{argmin}_{c \in C} |m_s - m_c|$. Figure 6.8 shows the resulting clusters for our example histogram and a threshold value of $T = 0.01$ with buckets corresponding to the same cluster being plotted with the same background.

6.3.3 Player-wise, Intersection-wise and Cluster-wise MC-Criticality

MC-criticality was suggested as an intuitive covariance measure between controlling a certain intersection of a Go board and winning the game by a number of people in 2009, e.g., in [29][79]. Here, controlling an intersection from the viewpoint of a specific player means, that in the game's terminal position, the intersection is either occupied by a stone of the player, or the intersection is otherwise counted as the players territory (cf. Section 2.1.1, page 8 for the exact meaning of the term *territory*). However, the complete information about which player p is controlling intersection i is only available in terminal positions. For non-terminal positions, we can try to estimate the information about control from MC simulations. Each MC simulation generates a terminal position and thereby a sample of control information for each single intersection. We can take each such sample as an observation of a random variable $X_{p,i}$ for each player-intersection pair, defined by

$$X_{p,i} = \begin{cases} 1, & \text{if player } p \text{ controls intersection } i \\ 0, & \text{else} \end{cases} .$$

In addition to the control information, each terminal position yields its associated position score. We use this score in the context of an additional random variable that we define for each of the score clusters that were found during the clustering procedure presented in Section 6.3.2:

$$X_c = \begin{cases} 1, & \text{if the score falls into cluster } c \\ 0, & \text{else} \end{cases} .$$

Hence, each MC simulation from a non-terminal position yields one terminal position and thereby a single observation of each of the above defined random variables. For example, for a 9×9 Go position, each MC simulation generates a single observation of $2 \cdot 81 + |C|$ random variables, where $|C|$ denotes the number of clusters found during clustering. Figure 6.9 depicts the observation of the random variables as described above.

As mentioned above, MC-criticality was before introduced as an intuitive covariance measure between controlling a certain intersection of a Go board and winning the game. We propose a slightly modified measure to compute the correlation of controlling a point on the Go board and achieving a score that falls into a given interval. Let $P = \{\text{Black}, \text{White}\}$ be the set of players, I the set of all board intersections and C a set of score clusters determined as presented in Section 6.3.2. We define the player-wise, intersection-wise and cluster-wise MC-criticality measure $g : P \times I \times C \rightarrow [-1, 1]$ by the correlation between the random variables $X_{p,i}$ and X_c:

$$g(p, i, c) := \text{Corr}(X_{p,i}, X_c) = \frac{\text{Cov}(X_{p,i}, X_c)}{\sqrt{\text{Var}(X_{p,i})}\sqrt{\text{Var}(X_c)}}$$

which gives

$$g(p, i, c) = \frac{\mu_{p,i,c} - \mu_{p,i}\mu_c}{Z} \quad \text{with } Z = \sqrt{\mu_{p,i} - \mu_{p,i}^2}\sqrt{\mu_c - \mu_c^2} \ ,$$

for $\mu_{p,i,c} = E[X_{p,i}X_c]$, $\mu_{p,i} = E[X_{p,i}]$ and $\mu_c = E[X_c]$ with $E[\cdot]$ denoting expectation. Accordingly, $\boldsymbol{\mu_{p,i,c}}$ denotes the ratio of all n simulations' terminal positions in which player p controlls intersection i and the score falls into cluster c, $\boldsymbol{\mu_{p,i}}$ represents the ratio of the simulations' terminal positions in which player p controls intersection i regardless of the score and $\boldsymbol{\mu_c}$ is the ratio of simulations with a final score that falls into cluster c.

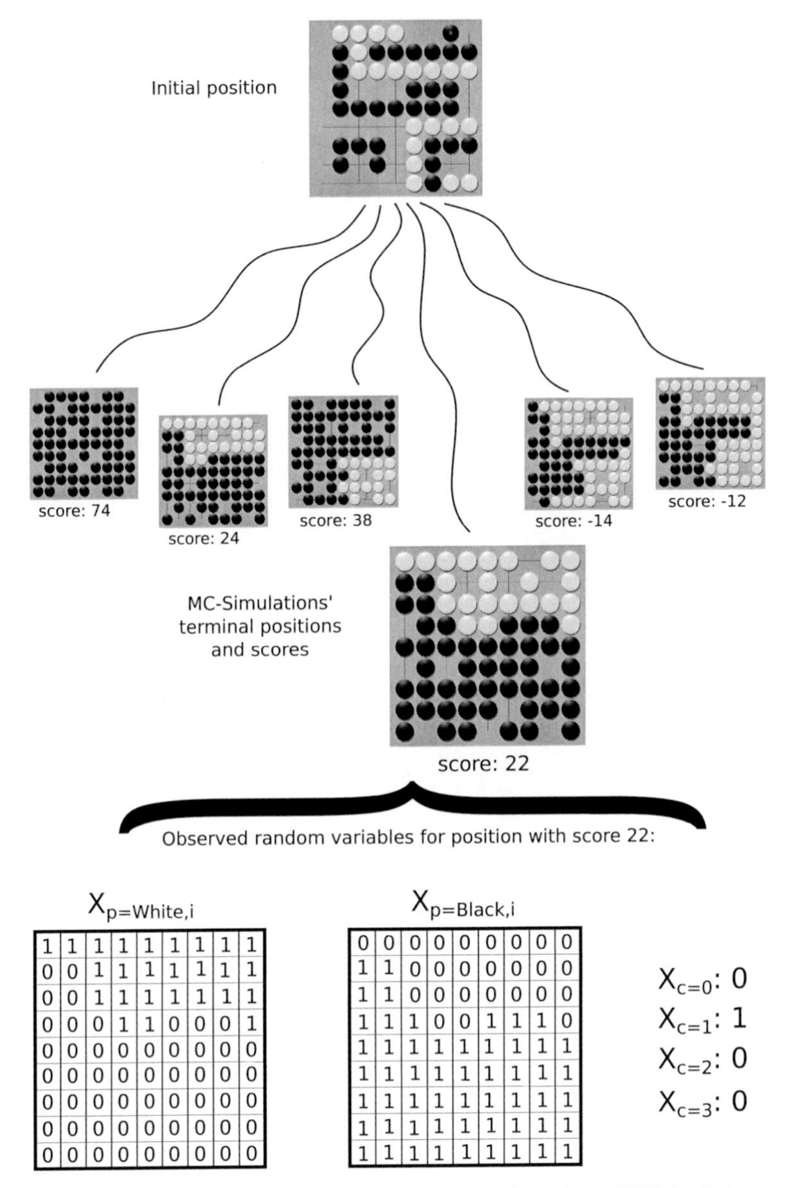

Figure 6.9: Observation of random variables at terminal positions of MC-simulations.

The measure $g(p,i,c)$ gives the criticality for player p to control intersection i by the end of the game in order to achieve some final score $s \in c$. The lowest possible value that indicates a complete negative correlation is -1. Here, negative correlation means that it is highly unlikely to end up in the desired score cluster if player p finally controls intersection i.

Our measure becomes most similar to the former published intersection-wise criticality measures when choosing the cluster $C_{\text{Black}} = \{s \in S | s > 0\}$ representing a Black win, and $C_{\text{White}} = \{s \in S | s < 0\}$ representing a White win. This clustering then resembles the criticality by

$$g_{\text{former}}(i) \approx g(\text{Black}, i, C_{\text{Black}}) + g(\text{White}, i, C_{\text{White}})$$

with the difference that $g(p,i,c)$ uses the correlation instead of the covariance.

6.3.4 Detecting and Identifying Local Fights

Putting the clustering procedure from Section 6.3.2 and the criticality measure of Section 6.3.3 together, we obtain a method for analyzing complete board positions with respect to a possible presence of local fights. In case a local fight exists, we are even able for approximately localizing it on the board.

To analyze a given position, we perform standard MCTS simulations and collect data about the simulations' terminal positions which is used to build the score histogram H and obtain values for $\mu_{p,i,c}$, μ_c and $\mu_{p,i}$. All we need for this purpose is a three dimensional array *control*, with a number of $|P| \cdot |I| \cdot |S|$ elements of a sufficiently large integer data type, initialize all elements to zero and increment them appropriately at the end of each MC simulation. Here, for each terminal position the value of element $\text{control}(p,i,s)$ is incremented in case player p controls intersection i and the position's score equals s. Using, e.g., 32-bit integers in 19×19 Go, the memory consumption would equal to $2 \cdot 361 \cdot 723 \cdot 4 = 2,085,136$ byte, hence about 2MB.

Given this, we create the histogram H by

$$H(s) = \sum_{p \in P} \text{control}(p,i,s) \quad \text{for any fixed } i \in I \ .$$

Note that any choice of i will lead to the same histogram, as we expect to see for each intersection, that exactly one of the two players is controlling it in terminal positions[5].

[5] Although expecting that each intersection is controlled by one of the two players is valid in general, there are the rare cases of seki points (cf. Section 2.1.1, page 8) in terminal positions that are controlled by neither player. Due to the rareness of seki we ignore this exceptions.

Having the score histogram of $n = \sum_{s \in S} H(s)$ simulations, we apply the clustering procedure as described in Section 6.3.2 to obtain the set of score clusters C. As mentioned in Sections 6.3.1 and 6.3.2, each cluster $c \in C$ is constructed around a mode of $\hat{f}$ and we denote the corresponding mode's position by m_c. Given this, we can derive the values for $\mu_{p,i,c}$, μ_c and $\mu_{p,i}$:

$$\mu_{p,i,c} = \frac{1}{n} \sum_{s \in c} \text{control}(p, i, s) \ ,$$

$$\mu_c = \frac{1}{n} \sum_{s \in c} H(s) \ ,$$

$$\mu_{p,i} = \frac{1}{n} \sum_{s \in S} \text{control}(p, i, s) \ .$$

This in turn allows for the cluster-wise criticality computation as described in Section 6.3.3. We hereby determine for each player the criticality of controlling an intersection in order to make the game end with a score belonging to the respective cluster. In case more than one cluster was found, the resulting distribution of criticality values for a given player and cluster, typically shows high valued regions that are object to a local fight. Thereby, the criticality values represent a stochastic mapping of each cluster to board regions with critical intersections that have to be controlled by one player in order to achieve a score that corresponds to the cluster. By further comparison of the critical board regions of the varying clusters and under consideration of the cluster's mode positions m_c, it might even be possible to estimate the value of a single fight in terms of scoring points.

As a special case it might happen for two or more score clusters corresponding to distinct regions on the board to overlap. Hereby, the score clusters might even become undistinguishable and accordingly appear as a single cluster only. In such special cases, the resulting single cluster would become mapped to the union of all involved critical regions.

In the next section, we present results achieved with our approach on a number of example Go positions.

6.3.5 Experiments

Based on our Go program Gomorra, we implemented our approach and made a series of experiments on different Go positions that contain multiple semeai. For our experiments, we concentrated on a number of two-safe-groups test cases out of a regression test suite created by Shih-Chieh (Aja) Huang (6 Dan) and Martin Müller [53]. The collection of problems in this test suite was especially created to reveal the weaknesses of current MCTS Go engines and is part of a larger regression test suite of the FUEGO Open Source project[6].

Figure 6.10 shows one of the test positions that contains two semeai, one on the upper right, the other on the lower right of the board. Black is to play and the result should be a clear win for White, hence a negative final score, because both semeai can be won by the white player. Figure 6.10a shows the corresponding score histogram of 128,000 MC simulations. The colors indicate the clustering computed by the method described in Section 6.3.2. Figures 6.10b, 6.10c, 6.10d and 6.10e show the respective criticality values for the white player to end up in cluster 1, 2, 3 and 4, counted from left to right. Positive correlations are illustrated by white squares of a size corresponding to the degree of correlation. Each intersection is additionally labeled with the criticality value. One can clearly see how the different clusters map to the different possible outcomes of the two semeai.

In the same manner, on page 116 Figures 6.11a to 6.11d show the results for test cases 1, 2, 3 and 5 of the above mentioned test suite (test case 4 is already shown in Figure 6.10). We restrict the result presentation to the score histograms and the criticality for player White to achieve a score assigned to the leftmost cluster. The histograms and criticality values always show the correct identification of two separate semeai. We took the leftmost cluster, as for the given test cases this is the one containing the correct evaluation, i.e., a win for the white player. As can be derived from the shown histograms, in all cases, Gomorra gets distracted by the other possible semeai outcomes and wrongly estimates the position as a win for Black as it is typical for MCTS programs that only work with the mean outcome of simulations.

Figure 6.12 shows results for a 13x13 game that was lost by the Go program Pachi playing Black against Alex Ketelaars (1 Dan). The game was played at the European Go Congress in Bonn in 2012 and was one of only two games lost by Pachi in the 13x13 tournament. As can be seen in the histogram, Alex managed to open up a number of fights. Again, the board shows the criticality values for the leftmost cluster and reveals Gomorra's difficulties in realizing that the lower left is clearly White's territory.

The position discussed above is one example of a number of games Ingo Althoefer collected on his website[7]. The site presents peak-rich histograms plotted by the Go program Crazy Stone. Althoefer came up calling such histograms *Crazy Shadows*. Rémi Coulom, the author of Crazy Stone, kindly generated a *Crazy Analysis* for the 13x13 game discussed above for comparison[8].

[6] Available online at `http://fuego.svn.sourceforge.net/viewvc/fuego/trunk/regression/`

[7] `http://www.althofer.de/crazy-shadows.html`

[8] `http://www.grappa.univ-lille3.fr/~coulom/CrazyStone/pachi13/index.html`

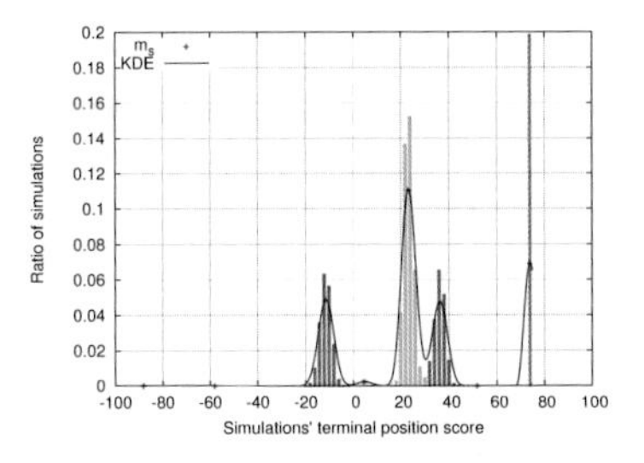

(a) Score histogram from 128,000 simulations.

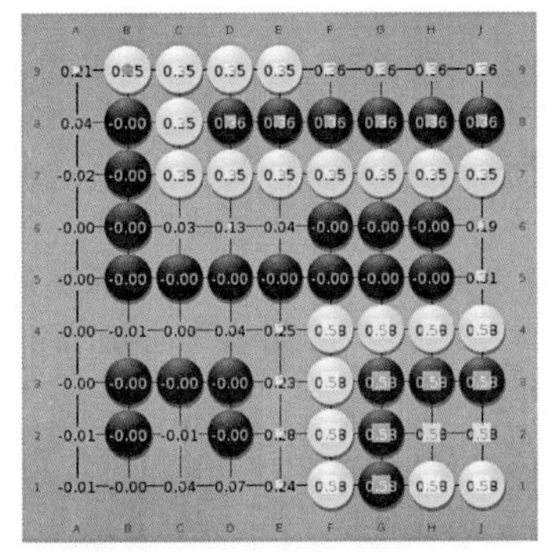

(b) Criticality for White to reach a score in cluster 1 (counted from left to right).

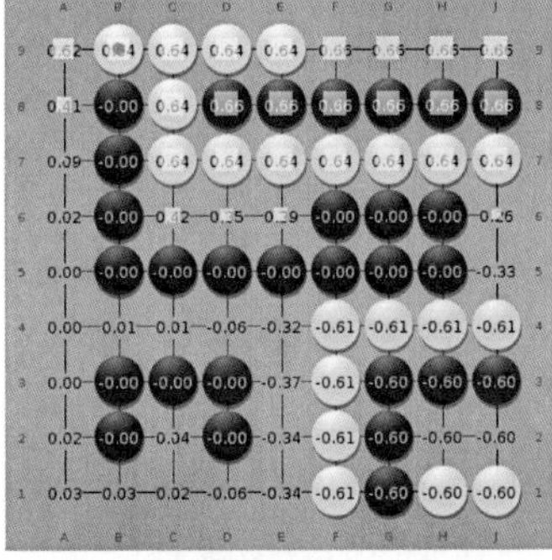

(c) White's criticality for cluster 2.

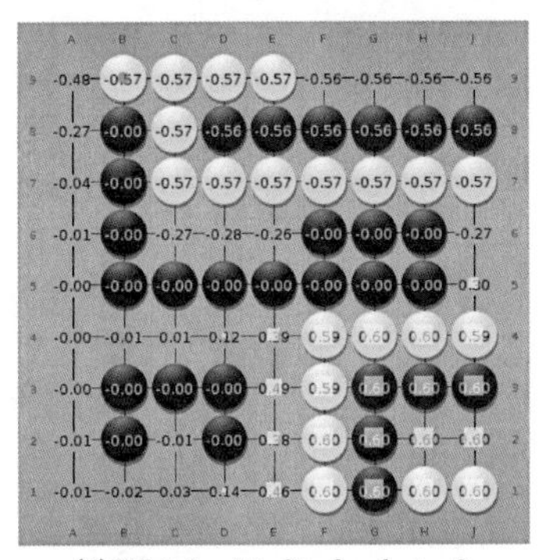

(d) White's criticality for cluster 3.

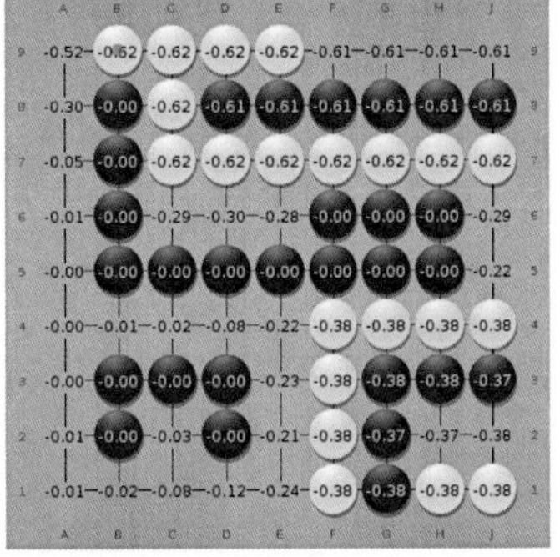

(e) White's criticality for cluster 4.

Figure 6.10: The player-wise criticality values reveal the critical intersections on the board for making the game end with a score associated to a specific cluster. The locations of the two capturing races are clearly identified.

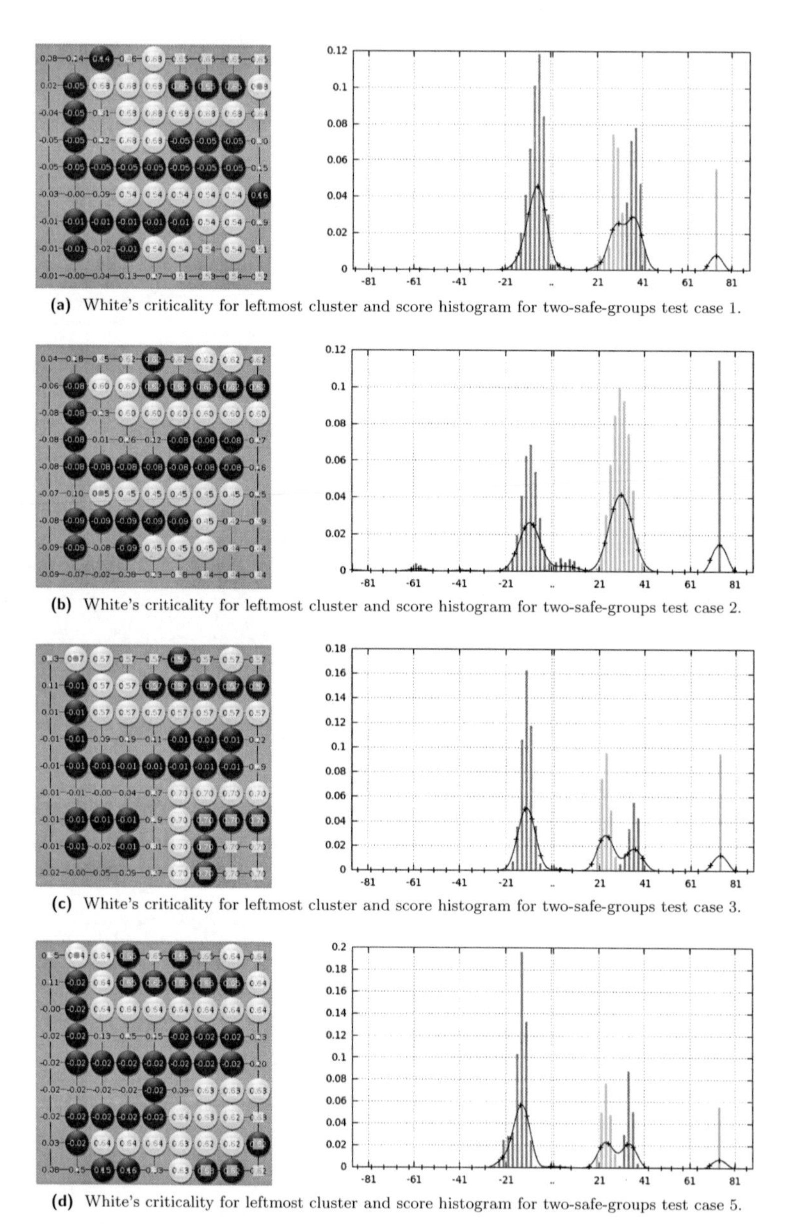

(a) White's criticality for leftmost cluster and score histogram for two-safe-groups test case 1.

(b) White's criticality for leftmost cluster and score histogram for two-safe-groups test case 2.

(c) White's criticality for leftmost cluster and score histogram for two-safe-groups test case 3.

(d) White's criticality for leftmost cluster and score histogram for two-safe-groups test case 5.

Figure 6.11: Player-wise criticality for another 4 test cases of the two-safe-groups test suite.

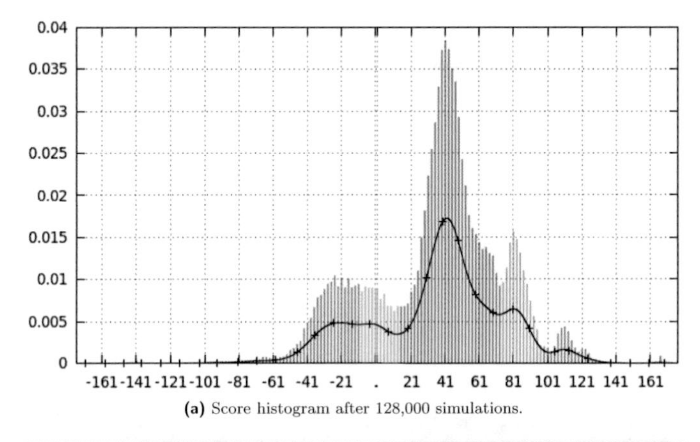

(a) Score histogram after 128,000 simulations.

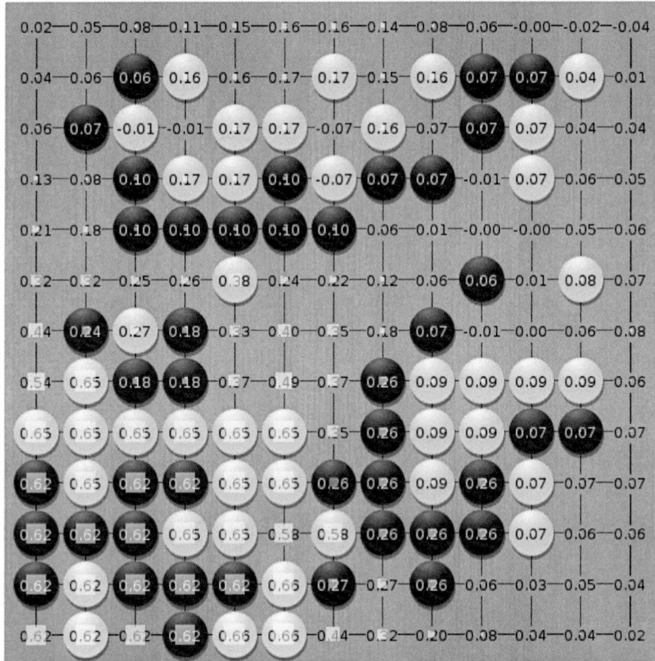

(b) Criticality for player White to reach a score of the leftmost cluster.

Figure 6.12: Analysis of a position occurred in the game between Go program Pachi (Black) vs. Alex Ketelaars at the European Go Congress 2012 in Bonn (move 90).

6.4 Chapter Conclusion

We presented a method to detect and localize local fights in Go positions and explained how to integrate the detection into existing MCTS implementations. Doing this in practice, we present a number of examples that demonstrate the power of our approach. However, the detection and localization of fights alone is only a first step towards improving the semeai handling capabilities of modern MCTS based Go programs. We must develop and evaluate methods to use the gathered knowledge in form of criticality values in the simulation policies to finally turn it into increased playing strength. Specifically, we are highly convinced that remarkable improvements can only be achieved when using the gathered information even in the playout policies. As the criticality computation is too time consuming to be done at every game state in the search tree we must restrict its computation to frequently visited tree nodes. It then might be well suited to support on-line learning techniques for move prediction models [104] to focus on critical board sites.

Accordingly, in future work, we plan to investigate the use of some kind of large asymmetric shape patterns that dynamically adapt their size and shape to the critical regions as they are determined by the presented method. Integrating those patterns into existing move prediction systems as they are widely used in Computer Go in addition with training their parameters during the search process builds the next interesting challenge. Already now, the results might be of interest for human Go players using Go programs to analyze game positions[9]. In chess, similar analysis tools were introduced in the early 1990's, for instance in the commercial database system by ChessBase company. They have become indispensable for masters' analysis.

[9]Some more context to existing visualizations for computer aided position analysis can be found online at `http://www.althoefer.de/k-best-visualisations.html`.

CHAPTER 7

Summary and Outlook

This chapter summarizes the contributions of this thesis and lists conclusions that can be drawn from our work. It further gives a number of potential future directions for the various topics discussed.

7.1 Contributions

Monte Carlo Tree Search is a very active field of research in these days. Starting with its enormous success in Computer Go, we regularly see new applications of MCTS to an ever increasing number of domains. In this thesis, we presented a new general technique for parallelizing MCTS on high performance compute clusters that could serve for faster and deeper searches in various search domains. We took the search domain of Computer Go for a practical implementation and empirical analysis of our approach, as it is the most studied MCTS domain by today. We further investigated the performance of move prediction systems and developed a new technique for an MCTS-guided automated position analysis for Go. In more detail, we made the following contributions to the research field of MCTS:

- We developed, implemented and analyzed a novel technique for parallelizing MCTS for hybrid shared and distributed memory systems. Our parallelization, called Distributed-Tree-Parallelization, maintains a single large game tree representation in memory. The distribution of the game tree allows for efficient usage of the entire distributed memory available in a cluster and thereby allows for more extensive recording of simulation statistics. We presented an efficient method for distributing a transposition table among the single compute nodes of a cluster that can safely process concurrent accesses. We further regulated the load balancing and introduced the use of additional compute nodes to leverage asynchronous all-to-all communication. We conducted extensive experiments that demonstrate the power and limitations of our approach. We empirically determined a scaling to more than 2000 compute cores in the best setting.

- As a base for the implementation of our parallelization, we developed an MCTS Go program called *Gomorra* with various state-of-the-art heuristics and extensions that yields a performance in terms of playing strength comparable to current strong Go programs. Implementing our parallelization on top of a strong state-of-the-art MCTS program makes us confident that our findings could be transferred to most other modern MCTS search implementations. Scaling to more than 1000 compute cores from a strong single node Go engine allowed us to get to the strength of the currently strongest Go programs in the world.
- Our comparison of a number of move prediction systems under equal conditions showed rather small differences in their respective prediction quality. Further analysis of the value of merging shape patterns with feature patterns gave rise to the assumption that further major improvements are possible. Partly motivated by our results, Wistuba et al. [105] developed a novel prediction system based on factorization machines that apparently outperforms the former best prediction systems.
- We believe that generalizing knowledge obtained from MCTS simulations is among the most promising future directions for Computer Go. MCTS based programs have particularly difficulties in evaluating positions when long sequences of tactical play are required for an accurate assessment. We developed a method for an automatic position analysis during regular MCTS searches. We thereby created a detection mechanism for potential evaluation inaccuracies and presented a way for heuristically localizing regions on the Go board that are responsible for such evaluation uncertainties.

7.2 Conclusions and Lessons Learned

From the work presented in this thesis, we draw the following conclusions:

While searching for a scalable parallelization for the Monte-Carlo Tree Search framework, we were guided from the idea of maintaining a single large distributed representation of the game tree in memory. This is in strong contrast to other parallelizations for distributed memory machines and allows for a more extensive collection of statistical sample data from MC simulations. Our empirical experimental results strengthen our conviction that this approach indeed leads to significant search quality improvements.

It was in doubt if MC simulations can efficiently be distributed on a large cluster, even if high performance networks are available. With the use of modern techniques like RDMA with a low latency Infiniband interconnect and efficient, concurrently accessible data structures, as well as load balancing techniques and a proper broadcast support mechanism, we were able to create a fully functional Go program that scales to more than a thousand compute cores.

But at the end, also the scalability of our approach is limited and we observe first degradations when using about 1000 compute cores. Our extensive experiments point to RDMA as an important source of our performance on the one hand but also a potential limiting technique when using many compute nodes on the other hand. RDMA requires for active polling of certain memory regions to perceive the completion of asynchronous communication. Such polling tasks bound CPU resources to an extent relative to the number of communication peers. This naturally goes along with increasing latencies and accordingly creates a scalability limitation when using many communication peers.

Certainly, also the ratio of MC simulations that are computed in total to the number of MC simulations that are computed in parallel is crucial for the obtainable overall performance. For example, performing 1000 simulations in total and performing all of them in parallel on 1000 compute units, we obviously end up in a result that cannot make use of the tree part of MCTS but is like a plain MC search. However, looking at a total number of 20M simulations, performing always 1000 in parallel might yield a similar search quality compared to computing all 20M simulations sequentially on a single compute unit. It is therefore especially important to ensure that the simulation rate (i.e., the number of simulations computable per time unit) scales well with the number of compute units and thereby the amount of parallel simulations. Given this, following our experimental results (cf. Figure 4.15, page 67), further strength improvements appear to be still realistic for systems with several thousand cores and several hundreds of compute nodes.

Our comparison of move prediction systems for the game of Go demonstrated the importance of the descriptive strength of the respective underlying probabilistic models. By aligning the single models regarding their expressiveness, we observed only minor differences in the prediction quality of either system, despite their rather diverse computational intensity.

Move prediction systems that are trained on Go games played by strong human players are beneficial for incorporating general knowledge about the specific search domain of Go. We are interested in detecting situations in that general knowledge is insufficient or even misleading. We developed a method for automatically detecting evaluation uncertainties that occur during MCTS searches that even allows for stochastically localizing corresponding regions on the Go board. The approach showed promising results for a number of sample positions and might help for augmenting move prediction for very specific situations in the future.

A proper field of research on games is combinatorial game theory (CGT). This area is about decomposing a game into subgames, solve such subgames independently to afterwards combine the single subgames' results to draw conclusions about the entire game. Related research on Computer Go (e.g., [77][76]) faces the problem of identifying subgames that can be solved independently. Our before mentioned method for MCTS guided position analysis might serve as an extractor for candidates of rather independent subgames in some cases. Note that CGT uses a proper calculus that differs from computing or estimating minmax values. Further integrating MCTS with CGT might be considered as a proper research field.

7.3 Future Directions

We will now list a number of promising future directions from the current state of our work:

- From our work on Distributed-Tree-Parallelization, a novel technique of parallelizing MCTS for hybrid shared and distributed memory systems, we learned that the use of a single large distributed representation of the game tree in memory is possible and allows for scaling to a large number of compute cores. The use of a single game tree representation is in contrast to another popular parallelization approach for distributed memory systems, the Root-Parallelization (cf. Section 4.1.1, page 39). Both approaches might benefit from different kinds of additional knowledge they obtain from additional compute resources. While Distributed-Tree-Parallelization allows for deeper searches by maintaining a single large tree representation, Root-Parallelization can benefit from several parallel, more independent MCTS searches that might prevent to get stuck in suboptimal branches of the search space. As numerous modern MCTS based Go programs give a strong bias towards exploitation when handling the exploration-exploitation tradeoff, mistakenly concentrating the search on suboptimal branches becomes more likely. This observation leads to the promising future direction of investigating potential combinations of both approaches, e.g., by performing several Distributed-Tree-Parallelization searches in parallel and connecting them via Root-Parallelization.

- The scalability limitations of Distributed-Tree-Parallelization are apparently induced by the RDMA technology or its kind of use in MPI distributions and our implementation. Accordingly, it appears worth investigating the use of other communication technologies. It might also be beneficial to improve the communication by exploiting further techniques available with the Infiniband interconnect, like its support for multicast operations. Doing this might be part of more general investigations on quantifying potential performance gains from handling network communication on lower levels, as an alternative to using MPI.
- A further critical part of Distributed-Tree-Parallelization as well as of Root-Parallelization is the synchronization of duplicated, frequently visited near root nodes. Developing even better and statistically sound policies for conducting node synchronization with the goal of minimizing communication overheads and ideally keeping them close to constant for single communication peers will likely improve the scalability of either parallelization approach.
- Distributed-Tree-Parallelization is so far only implemented for Computer Go. Although we looked for keeping the parallelization itself as general as possible, it is considered worthwhile to extend empirical experiments to further search domains. Hence, implementing Distributed-Tree-Parallelization for other search domains poses another future direction.
- Our technique for detecting and localizing evaluation uncertainties in Go positions during MCTS searches appears to be well suited to be used for detecting further special phenomena that show up in games. In case of Go, e.g., the same method should be adaptable to detect and localize potential seki points. Hence, a more general tool for MCTS guided analysis of Go positions might be obtainable.
- So far, our aforementioned approach is able to detect evaluation uncertainties and can be used to identify critical regions in Go positions. For a particular position, it should be possible to develop additional tiny local MCTS search trees for each of such critical regions. Those could be faded in the principal tree at corresponding places. The fade in might be realized similar to the incorporation of AMAF statistics in the RAVE extension (cf. Section 3.3.1). Furthermore, statistics stored with such local trees could be used for guiding playouts in critical regions, thereby making them adaptive to local, critical situations. We obtained promising results with the latter approach in early experiments.
- It is often mentioned that the use of accelerator hardware like FPGAs or GPGPUs should have a great potential for being used in the context of MCTS. As currently used playout policies are however far from uniformly

random, they do not offer many places that appear amenable for fine grained parallelization. As a result, despite many attempts, none of the currently strongest Go playing programs make use of accelerator hardware. A promising field for future research might be attempts to learn locally generalizable knowledge from the data produced by MC simulations online, i.e., during the actual MCTS search. This might involve common techniques of data mining and machine learning like neural networks that can greatly benefit from accelerator hardware already today. Hence, conducting research in this field with the opportunities of accelerator hardware in mind might lead to a next major step in Computer Go and related areas.

Bibliography

[1] Bruce Abramson. *The Expected-Outcome Model of Two-Player Games.* PhD thesis, Columbia University, 1987.

[2] Bruce Abramson. An analysis of expected-outcome. *Journal of Experimental and Theoretical Artificial Intelligence*, 2:55–73, 1990.

[3] Ingo Althöfer. On pathology in game tree and other recursion tree models. Habilitation Thesis, Faculty of Mathematics, University of Bielefeld, June 1991.

[4] Nobuo Araki, Kazuhiro Yoshida, Yoshimasa Tsuruoka, and Jun'ichi Tsujii. Move Prediction in Go with the Maximum Entropy Method. In *IEEE Symposium on Computational Intelligence and Games*, pages 189–195, April 2007.

[5] Broderick Arneson, Ryan B. Hayward, and Philip Henderson. Monte Carlo Tree Search in Hex. *IEEE Transactions on Computational Intelligence and AI in Games*, 2(4):251–258, December 2010.

[6] Peter Auer, Nicolò Cesa-Bianchi, and Paul Fischer. Finite-Time Analysis of the Multiarmed Bandit Problem. In *Machine Learning*, volume 47, pages 235–256. Kluwer Academic, 2002.

[7] Hendrik Baier and Peter D. Drake. The Power of Forgetting: Improving the Last-Good-Reply Policy in Monte Carlo Go. In *IEEE Transactions on Computational Intelligence and AI in Games*, pages 303–309, 2010.

[8] Petr Baudis. Information Sharing in MCTS. Master's thesis, Charles University in Prague, 2011.

[9] Petr Baudis and Jean-Loup Gailly. PACHI: State of the Art Open Source Go Program. In H. Jaap van den Herik and Aske Plaat, editors, *Advances in Computer Games 13*, volume 7168 of *LNCS*, pages 24–38. Springer, 2011.

[10] R. C. Bell. *BOARD AND TABLE GAMES from Many Civilizations, Chapter 5: Territorial Possession.* 1979.

[11] D. Benson. Life in the game of Go. *Inform. Sci.*, 10:17–29, 1976.

[12] M. Boon. A pattern matcher for Goliath. *Computer Go*, pages 13–23, 1990.

[13] Amine Bourki, Guillaume M.J-B. Chaslot, Matthieu Coulm, Vincent Danjean, Hassen Doghmen, Jean-Baptiste Hoock, Thomas Hérault, Arpad Rimmel, Fabien Teytaud, Olivier Teytaud, Paul Vayssiére, and Ziqin Yu. Scalability and Parallelization of Monte-Carlo Tree Search. In *International Conference on Computers and Games*, pages 48–58, 2010.

[14] Bruno Bouzy. History and Territory Heuristics for Monte-Carlo Go. *New Mathematics and Natural Computation*, 2(2):1–8, 2006.

[15] Bruno Bouzy and Tristan Cazenave. Computer Go: An AI oriented survey. *Artificial Intelligence*, 132(1):39–103, October 2001.

[16] Bruno Bouzy and Guillaume M.J-B. Chaslot. Monte-Carlo Go Reinforcement Learning Experiments. In *IEEE Symp. on Comp. Intelligence and Games*, pages 187–194, 2006.

[17] Bruno Bouzy and Bernard Helmstetter. Monte Carlo Go Developments. In *10th Advances in Computer Games Conference*, pages 159–174, 2004.

[18] Ralph Allan Bradley and Milton E. Terry. Rank Analysis of Incomplete Block Designs: I. The Method of Paired Comparisons. *Biometrika*, 39(3/4):324–345, December 1952.

[19] Cameron Browne, Edward Powley, Daniel Whitehouse, Simon Lucas, Peter I. Cowling, Philipp Rohlfshagen, Stephen Tavener, Diego Perez, Spyridon Samothrakis, and Simon Colton. A Survey of Monte Carlo Tree Search Methods. *IEEE Transactions on Computational Intelligence and AI in Games*, 4(1):1–43, March 2012.

[20] Bernd Brügmann. Monte Carlo Go. , 1993.

[21] Tristan Cazenave and Nicolas Jouandeau. On the Parallelization of UCT. In *ICGA Computer Games Workshop*, pages 93–101, June 2007.

[22] Tristan Cazenave and Nicolas Jouandeau. A Parallel Monte-Carlo Tree Search Algorithm. In *International Conference on Computer and Games*, volume 5131 of *LNCS*, pages 72–80, 2008.

[23] Guillaume M.J-B. Chaslot, Louis Chatriot, C. Fiter, Sylvain Gelly, Jean-Baptiste Hoock, Julien Perez, Arpad Rimmel, and Olivier Teytaud. Combining expert, offline, transient and online knowledge in Monte-Carlo explo-

ration. In *European Workshop on Reinforcement Learning*, 2008.

[24] Guillaume M.J-B. Chaslot, Mark H.M. Winands, and H. Jaap van den Herik. Parallel Monte-Carlo Tree Search. In *Conference on Computers and Games*, pages 60–71, 2008.

[25] Guillaume M.J-B. Chaslot, Mark H.M. Winands, H. Jaap van den Herik, and Jos W.H.M. Uiterwijk. Progressive Strategies for Monte-Carlo Tree Search. *New Mathematics and Natural Computation*, 4(3):343–357, November 2008.

[26] Yizong Cheng. Mean Shift, Mode Seeking, and Clustering. *IEEE Transactions on Pattern Analysis and Machine Intelligence*, 17(8):790–799, August 1995.

[27] Rémi Coulom. Efficient Selectivity and Backup Operators in Monte-Carlo Tree Search. In *Int. Conf. on Computers and Games*, volume 4630 of *LNCS*, pages 72–83. Springer, 2006.

[28] Rémi Coulom. Computing Elo Ratings of Move Patterns in the Game of Go. In *ICGA Journal*, volume 30, pages 198–208, 2007.

[29] Rémi Coulom. Criticality: a Monte-Carlo Heuristic for Go Programs. Invited talk at the University of Electro-Communications, Tokyo, Japan, January 2009.

[30] Benjamin Doerr and Ulf Lorenz. Error Propagation in Game Trees. *Mathematical Methods of Operations Research*, 64(1):79–93, August 2006.

[31] Chrilly Donninger, Alex Kure, and Ulf Lorenz. Parallel Brutus: The First Distributed, FPGA Accelerated Chess Program. In *18th International Parallel and Distributed Processing Symposium*. IEEE Computer Society, April 2004.

[32] Chrilly Donninger and Ulf Lorenz. The Chess Monster Hydra. In *Proc. of Int. Conf. on Field-Programmable Logic and Applications (FPL)*, volume 3203 of *LNCS*, pages 927–932, 2004.

[33] Peter D. Drake. The Last-Good-Reply Policy for Monte-Carlo Go. *ICGA Journal*, 32(4):221–227, December 2009.

[34] Arpad E. Elo. *The Rating of Chessplayers, Past and Present.* Arco Publishing, New York, 1986.

[35] Markus Enzenberger and Martin Müller. A Lock-free Multithreaded Monte-Carlo Tree Search. In *12th International Conference on Advances in Computer Games*, volume 6048 of *LNCS*, pages 14–20. Springer-Verlag, May 2009.

[36] Rainer Feldmann, Peter Mysliwietz, and Burkhard Monien. Distributed Game Tree Search on a Massively Parallel System. In *Data Structures and Efficient Algorithms*, volume 594 of *LNCS*, pages 270–288. Springer-Verlag, 1992.

[37] Hilmar Finnsson and Yngvi Björnsson. Simulation-Based Approach to General Game Playing. In *Proc. of the 23. AAAI Conf. on Artificial Intelligence*, pages 259–264, 2008.

[38] David Fotland. Knowledge representation in The Many Faces of Go. In *2nd Cannes/Sophia-Antipolis Go Workshop*, 1993.

[39] Keinosuke Fukunaga and Larry D. Hostetler. The Estimation of the Gradient of a Density Function, with Applications in Pattern Recognition . *IEEE Transactions on Information Theory*, 21(1):32–40, January 1975.

[40] Haiying Gao, Fuming Wang, Wei Lei, and Yun Lin. Monte Carlo simulation of 9x9 Go game on FPGA . In *IEEE International Conference on Intelligent Computing and Intelligent Systems*, pages 865–869, October 2010.

[41] R. Gaudel and M. Sebag. Feature Selection as a One-Player Game. In *Proc. 27th Int. Conf. Mach. Learn.*, pages 359–366, 2010.

[42] Sylvain Gelly and David Silver. Combining Online and Offline Knowledge in UCT. In *International Conference on Machine Learning*, pages 273–280, 2007.

[43] Sylvain Gelly, Yizao Wang, Remi Munos, and Olivier Teytaud. Modifications of UCT with Patterns in Monte-Carlo Go. Technical Report 6062, INRIA, 2006.

[44] Thore Graepel, Mike Goutrie, Marco Krüger, and Ralf Herbrich. Learning on Graphs in the Game of Go. In *Proc. of the Ninth Int. Conf. on Artificial Neural Networks*, January 2001.

[45] Tobias Graf, Ulf Lorenz, Marco Platzner, and Lars Schaefers. Parallel Monte-Carlo Tree Search for HPC Systems. In *Proceedings of the 17th International Conference, Euro-Par*, volume 6853 of *LNCS*, pages 365–376. Springer, Heidelberg, August 2011.

[46] Tobias Graf, Lars Schaefers, and Marco Platzner. On Semeai Detection in Monte-Carlo Go. In *Int. Conf. on Computers and Games*, 2013.

[47] John C. Handley. Comparative Analysis of Bradley-Terry and Thurstone-Mosteller Paired Comparison Models for Image Quality Assessment. In *Proc. of the IS and TS PICS Conf.*, pages 108–112, 2001.

[48] David P. Helmbold and Aleatha Parker-Wood. All-Moves-As-First Heuristics in Monte-Carlo Go. In *Int. Conf. on Artificial Intelligence (ICAI)*, pages 605–610, 2009.

[49] Ralf Herbrich, Tom Minka, and Thore Graepel. TrueSkill(TM): A Bayesian skill rating system. In *Advances in Neural Information Processing Systems 20*, pages 569–576. MIT Press, 2007.

[50] Kai Himstedt, Ulf Lorenz, and Dietmar P. F. Möller. A Twofold Distributed Game-Tree Search Approach Using Interconnected Clusters. In *Euro-Par*, volume 5168 of *LNCS*, pages 587–598. Springer, 2008.

[51] Shih-Chieh Huang, Rémi Coulom, and Shun-Shii Lin. Monte-Carlo Simulation Balancing in Practice. In *Int. Conf. on Computers and Games*, pages 81–92, 2010.

[52] Shih-Chieh Huang, Rémi Coulom, and Shun-Shii Lin. Time Management for Monte-Carlo Tree Search Applied to the Game of Go. In *International Conference on Technologies and Applications of Artificial Intelligence (TAAI)*, pages 462–466, 2010.

[53] Shih-Chieh Huang and Martin Müller. Investigating the Limits of Monte Carlo Tree Search Methods in Computer Go. In *International Conference on Computers and Games*, LNCS. Springer, 2013.

[54] David R. Hunter. MM algorithms for generalized Bradley Terry models. *The Annals of Statistics*, 32(1):384–406, 2004.

[55] Robert Hyatt and Tim Mann. A lock-less transposition table implementation for parallel search chess engines. *ICGA Journal*, 25(1), 2002.

[56] Y. K. Kao. *Sum of hot and tepid combinatorial games.* PhD thesis, Univerity of North Carolina at Charlotte, 1997.

[57] Hideki Kato and Ikuo Takeuchi. Parallel Monte-Carlo Tree Search with Simulation Servers. In *Proc. of the Int. Conf. on Technologies and Applications of Artificial Intelligence*, pages 491–498, 2010.

[58] Akihiro Kishimoto and Jonathan Schaeffer. Transposition Table Driven Work Scheduling in Distributed Game-Tree Search. In *Canadian Conference on Artificial Intelligence*, volume 2338 of *LNCS*, pages 56–68. Springer Berlin/Heidelberg, 2002.

[59] E. Donald Knuth and Ronald W. Moore. An Analysis of Alpha-Beta Pruning. In *Artificial Intelligence*, volume 6, pages 293–327. North-Holland Publishing Company, 1975.

[60] Levente Kocsis and Csaba Szepesvári. Bandit based Monte-Carlo Planning. In *ECML*, volume 4212 of *LNCS/LNAI*, pages 282–293, 2006.

[61] T. Lai and H. Robbins. Asymptotically efficient adaptive allocation rules. *Advances in Applied Mathematics*, 6:4–22, 1985.

[62] D. Lefkovitz. A strategic pattern recognition program for the game of Go. Technical Report 60-243, 1–92, University of Pennsylvania, The Moore School of Electrical Engineering, Wright Air Development Division, 1960.

[63] Gottfried Wilhelm Leibniz. Annotatio de quibusdam Ludis; inprimis de Ludo quodam Sinico, differentiaque Scachici & Latrundulorum, & novo genere Ludi Navalis, 1710.

[64] Łukasz Lew. *Modeling Go Game as a Large Decomposable Decision Process.* PhD thesis, Warsaw University, June 2011.

[65] David Lichtenstein and Michael Sipser. GO is Polynomial-Space Hard. *Journal of Association for Computing Machinery*, 27(2):393–401, April 1980.

[66] Richard J. Lorentz. Amazons Discover Monte-Carlo. In *Int. Conf. on Computers and Games*, volume 5131 of *LNCS*, pages 13–24, 2008.

[67] Richard J. Lorentz and Therese Horey. Programming Breakthrough. In *Int. Conf. on Computers and Games*, 2013.

[68] Ulf Lorenz. *Controlled Conspiracy Number Search.* PhD thesis, University of Paderborn, December 2000.

[69] Ulf Lorenz. Parallel Controlled Conspiracy Number Search. In *Proc. of Int. Euro-Par Conf. (Euro-Par)*, volume 2400 of *LNCS*, pages 420–430, 2002.

[70] Ulf Lorenz. Beyond Optimal Play in Two-Person-Zerosum Games. In S. Albers and T. Radzik, editors, *Proc. of the 12th Annual European Symposium on Algorithms (ESA)*, volume 3221 of *LNCS*, pages 749–759, 2004.

[71] Ulf Lorenz and Burkhard Monien. Error Analysis in Minimax Trees. *Journal of Theoretical Computer Science (TCS)*, 313(3):485–498, February 2004.

[72] David J. C. MacKay. *Information Theory, Inference, and Learning Algorithms*, chapter IV-26: Exact Marginalization in Graphs. Cambridge University Press, 2003.

[73] T. Mahlmann, J. Togelius, and G. N Yannakakis. Towards Procedural Strategy Game Generation: Evolving Complementary Unit Types. In *Proc. Applicat. Evol. Comput.* , volume 6624 of *LNCS*, pages 93–102, 2011.

[74] David McAllester. Conspiracy Numbers for Min-Max Search. *Artificial Intelligence*, 35(1):287–310, 1988.

[75] Tom Minka. *A family of algorithms for approximate Bayesian inference.* PhD thesis, Massachusettes Institute of Technology, 2001.

[76] Martin Müller. In *Int. Joint Conf. on Artificial Intelligence (IJCAI-99)*, pages 578–583.

[77] Martin Müller. *Computer Go as a sum of local games: An application of combinatorial game theory.* PhD thesis, Swiss Federal Institute of Technology, 1995.

[78] H. Nakhost and Martin Müller. Monte-Carlo Exploration for Deterministic Planning. In *Proc. 21st Int. Joint Conf. Artif. Intell.*, pages 1766–1771, 2009.

[79] Seth Pellegrino, Andrew Hubbard, Jason Galbraith, Peter D. Drake, and Yung-Pin Chen. Localizing Search in Monte-Carlo Go Using Statistical Covariance. *ICGA Journal*, 32(3):154–160, 2009.

[80] H. Remus. Simulation of a learning machine for playing Go. In *Proc. of the IFIP (Int. Federation of Information Processing)*, pages 428–432, 1963.

[81] Arpad Rimmel, Olivier Teytaud, Chang-Shing Lee, Shi-Jim Yen, Mei-Hui Wang, and Shang-Rong Tsai. Current Frontiers in Computer Go. In *IEEE Transactions on Computational Intelligence and AI in Games*, volume 2, pages 229–238, December 2010.

[82] J. M. Robson. Complexity of Go. In *Proc. of the IFIP (Int. Federation of Information Processing)*, pages 413–417, 1983.

[83] Kamil Rocki and Reiji Suda. Large-Scale Parallel Monte Carlo Tree Search on GPU. In *IEEE International Symposium on Parallel and Distributed Processing Workshops and Phd Forum*, pages 2034–2037, May 2011.

[84] Kamil Marek Rocki. *Large Scale Monte Carlo Tree Search on GPU.* PhD thesis, University of Tokyo, August 2011.

[85] John W. Romein, Aske Plaat, Henri E. Bal, and Jonathan Schaeffer. Transposition Table Driven Work Scheduling in Distributed Search. In *National Conference on Artificial Intelligence*, pages 725–731, 1999.

[86] B. Satomi, Y. Joe, A. Iwasaki, and M. Yokoo. Real-Time Solving of Quantified CSPs Based on Monte-Carlo Game Tree Search. In *Proc. 22nd Int. Joint Conf. Artif. Intell.*, pages 655–662, 2011.

[87] Lars Schaefers and Marco Platzner. UCT-Treesplit - Parallel MCTS on Distributed Memory. In *MCTS Workshop at Int. Conf. on Automated Planning and Scheduling (ICAPS)*, June 2011.

[88] Günther Schrüfer. Presence and Absence of Pathology on Game Trees. In *Advances in computer chess*, pages 101–112, Elmsford, NY, USA, 1986. Pergamon Press, Inc.

[89] Richard B. Segal. On the Scalability of Parallel UCT. In *International Conference on Computer and Games*, pages 36–47, 2010.

[90] Claude E. Shannon. Programming a Computer for Playing Chess. *Philosophical Magazin*, 41(314):256–275, March 1950. First presented at the National IRE Convention, March 9, 1949, New York, U.S.A.

[91] David Silver. *Reinforcement Learning and Simulation-Based Search in Computer Go*. PhD thesis, University of Alberta, 2009.

[92] David Silver, Richard Sutton, and Martin Müller. Sample-based learning and search with permanent and transient memories. In *Proceedings of the 25th International Conference on Machine Learning*, pages 968–975, 2008.

[93] David Silver and Gerald Tesauro. Monte-Carlo Simulation Balancing. In *International Conference on Machine Learning*, pages 945–952, 2009.

[94] Bernard W. Silverman. *Density Estimation for Statistics and Data Analysis*, volume 26 of *Monographs on Statistics and Applied Probability*. Chapman & Hall/CRC, April 1986.

[95] David Stern. *Modelling Uncertainty in the Game of Go*. PhD thesis, University of Cambridge, February 2008.

[96] David Stern, Ralf Herbrich, and Thore Graepel. Bayesian Pattern Ranking for Move Prediction in the Game of Go. In *Proceedings of the International Conference of Machine Learning*, January 2006.

[97] David Stoutamire. Machine learning, game play, and Go. Technical Report 91-128, Case Western Reserve University, 1991.

[98] Fabien Teytaud and Olivier Teytaud. On the Huge Benefit of Decisive Moves in Monte-Carlo Tree Search Algorithms. In *IEEE Symposium on Computational Intelligence and Games*, number 1, pages 359–364, 2010.

[99] E. Thorpe and W. Walden. A computer assisted study of Go on M x N boards. *Inform. Sci.*, 4:1–33, 1972.

[100] John Tromp and Gunnar Farnebäck. Combinatorics of Go. In *Proc. of the Int. Conf. on Computers and Games*, pages 84–99, Berlin, Heidelberg, 2007.

Springer-Verlag.

[101] Erik van der Werf, Jos W.H.M. Uiterwijk, Eric Postma, and H. Jaap van den Herik. Local Move Prediction in Go. In *Conference on Computers and Games*, volume 2883 of *LNCS*, pages 393–412, 2003.

[102] Erik van der Werf, H. Jaap van den Herik, and Jos W.H.M. Uiterwijk. Solving Go on Small Boards. *ICGA Journal*, 26(2):92–107, 2003.

[103] Ruby C. Weng and Chih-Jen Lin. A Bayesian Approximation Method for Online Ranking. *Journal of Machine Learning Research*, 12:267–300, January 2011.

[104] Martin Wistuba, Lars Schaefers, and Marco Platzner. Comparison of Bayesian Move Prediction Systems for Computer Go. In *Proc. of the IEEE Conf. on Computational Intelligence and Games (CIG)*, pages 91–99, September 2012.

[105] Martin Wistuba and Lars Schmidt-Thiele. Move Prediction in Go – Modelling Feature Interactions Using Latent Factors. In *KI2013 - German Conference on Artificial Intelligence*, pages 260–271. Springer Berlin/Heidelberg, 2013.

[106] Thomas Wolf. The program GoTools and its computer-generated tsume-go database. In *Proc. of the First Game Programming Workshop in Japan*, page 84, 1994.

[107] Kazuki Yoshizoe, Akihiro Kishimoto, Tomoyuki Kaneko, Haruhiro Yoshimoto, and Yutaka Ishikawa. Scalable Distributed Monte-Carlo Tree Search. In Daniel Borrajo, Maxim Likhachev, and Carlos Linares López, editors, *SOCS*, pages 180–187. AAAI Press, 2011.

[108] Albert Lindsey Zobrist. A model of visual organization for the game of Go. In *Proc. of the AFIPS Spring Joint Computer Conf.*, pages 103–111, 1969.

[109] Albert Lindsey Zobrist. A New Hashing Method with Application for Game Playing. Technical Report 88, Computer Sciences Department, University of Wisconsin, 1969.

[110] Albert Lindsey Zobrist. *Feature Extraction and Representation for Pattern Recognition and the Game of Go.* PhD thesis, University of Wisconsin, August 1970.